D1369932

LECTURE NOTES ON
Bacteriology

R. R. GILLIES
M.D., D.P.H., M.C.Path.

Senior Lecturer in Bacteriology
University of Edinburgh Medical School

BLACKWELL SCIENTIFIC PUBLICATIONS
OXFORD AND EDINBURGH

© BLACKWELL SCIENTIFIC PUBLICATIONS 1968

*This book is copyright. It may not be reproduced by any means
in whole or in part without permission. Application with regard
to copyright should be addressed to the publishers.*

SBN 632 01860 7

FIRST PUBLISHED 1968

Printed in Great Britain by
WESTERN PRINTING SERVICES LTD, BRISTOL
and bound by
THE KEMP HALL BINDERY, OXFORD

Contents

Preface

This small volume has been prepared in the hope that, in at least some of his lectures in bacteriology, the student will be saved the task of scribbling notes; the text is not meant to be exhaustive as will be obvious from the size of the volume in comparison with that of many other textbooks on the subject. An attempt has been made to highlight those features of bacterial species which are important in their identification and in the laboratory diagnosis of infection. Additionally, brief notes have been included on the epidemiology and prevention of certain infections.

CHAPTER 1
Historical Introduction

The history of any science usually attracts interest from the older practitioners of that science and from professional historians; nevertheless in the history of bacteriology there is much to stimulate all bacteriologists, including young students of the subject.

It is, for example, fascinating to realize that centuries before bacteria had been discovered, several authorities had postulated their existence and written detailed accounts of how they spread within communities and in some instances had advocated measures to prevent the spread of diseases which we now know to have a bacterial origin.

There has been much specialization within the science of bacteriology in the last few decades; the *fundamental bacteriologist*, who is more often a science graduate than a medical graduate, studies bacteria for their own sake; and researches on, for example, bacterial morphology, biochemistry or genetics are not primarily undertaken with any medical application in view. Antonie van Leeuwenhoek (1632–1723) may be regarded as the father of fundamental microbiology since, although Kircher (1602–80) may have been the first to see bacteria, van Leeuwenhoek (1676) carried out the earliest recorded investigation of bacteria in several environments, but his interests in his 'little animals' were restricted mainly to their natural inanimate habitats.

Other bacteriologists specialize in the study of the interaction of bacteria and their human and animal hosts and are termed *epidemiologists*. Epidemiology as a science now encompasses many fields in addition to the study of the sources and methods of spread of communicable diseases, but the basic practice is identical whether one is studying communicable diseases, non-communicable diseases (such as peptic ulcer, diabetes or lung cancer), or even studying the epidemiology of traffic accidents. In each instance the epidemiologist attempts to relate environmental causes to the ultimate effect with the aim of preventing the disease by interrupting the chain of events. Fracastòrius (1478–1553) may be regarded as the earliest epidemiologist and indeed we still use certain of his words and phrases in present-day epidemiological thinking and writing; for example, he described clearly the part played by fomites—inanimate objects—in allowing the transfer of contagion from one person to another.

Bacteriology made its debut as a result of the intensive investigations carried out by Koch (1843–1910) and Pasteur (1822–95) and the *clinical bacteriologist* then came into being; unlike *fundamental* bacteriology, which is involved solely with the bacterial cell, or *epidemiology*, which concentrates on community phenomena associated with bacteria, clinical bacteriology focuses the attention at the personal level of the individual patient who is infected and the clinical bacteriologist is intimately involved—or should be —in determining the most suitable antimicrobial agent to treat the sick person. Paul Ehrlich's dream of the magic bullet, i.e. the use of a single agent which could be used to treat all bacterial infections, has almost been realized since the introduction of sulphonamides in 1935 and penicillin (1942), the forerunner of the many antibiotics now available. It is important to realize that the blind use of such powerful antimicrobial agents carries many hazards, not only for the patient but also for the community. A patient not suffering from bacterial infection may be 'treated' with such a drug without benefit; another patient may be suffering from a bacterial infection and be given an antibiotic to which the infecting organism is resistant. Even when laboratory identification of the organism has been made and its *in-vitro* pattern of sensitivity to various agents established, many antimicrobial drugs carry a risk of side-effects, some of which may be so severe, that one could not justify their use in some cases. The risk to the community, even when such agents are used intelligently, lies mainly in bacteria acquiring resistance to one or more antimicrobial drugs so that resistant strains may spread and cause disease against which a diminishing number of antibiotics is available.

The clinical bacteriologist has a very important role in the monitoring of special units in the modern hospital; certain patients are at very great risk if they become infected, e.g. the individual requiring kidney transplantation is treated with certain agents to depress his immune response mechanisms to facilitate the acceptance of the grafted kidney by his tissues. These same agents therefore increase the patient's susceptibility to infection and his environment must be as free from bacteria as possible; not only does the bacteriologist participate in the design of such patient-care areas but he has the responsibility of constantly checking on their safety. He is equally involved in the intensive-care areas such as assisted respiration units and neonatal nurseries. On a wider front the bacteriologist is involved in reducing the risk of infection from instruments and dressings by checking Central Sterile Supply Departments and Theatre Service Centres. The latter concept aims at supplying operating theatres with instruments and other materials which are *sterile* and provided in a fashion convenient for each surgeon; in addition such centralization removes the physically tiring pre- and post-operative preparation of materials for the theatre nursing staff.

The image of the bacteriologist, held by many students and too many practitioners, as an individual who blindly examines and reports on material submitted to him is slowly dying; the clinical bacteriologist is fully committed to all aspects of patient-care, including the prevention of infection.

CHAPTER 2
Bacterial Anatomy

Studies of the anatomy of the bacterial cell appear in many erudite publications; however, in this chapter we must concentrate on the features which have a practical application in medical bacteriology.

Figure I illustrates the essential features of a 'typical' bacterial cell.

Cell Wall and Cytoplasmic Membrane

Almost all bacteria possess a cell wall which is sufficiently strong to give each cell a particular shape and its relationship to the underlying *cytoplasmic membrane* has been likened to that which the outer cover of a pneumatic tyre has to the inner tube. The mechanical strength of the cell wall is evident when, under strictly controlled circumstances, it is removed to release the remainder of the cell—a free protoplast—in an intact state. Regardless of the original shape of the cell, the protoplast is spherical and without the protection of the cell wall is osmotically sensitive, so that if there is variation in the osmotic pressure of the fluid in which the protoplast is maintained it decreases or increases in size. Although protoplasts will continue to function as did the intact cell, they do not subdivide.

(The cell wall participates in cell division and in determining the spatial relationship of organisms) In the formation of daughter cells, and soon after the nuclear material has undergone fission, the cell wall grows in at the equator of the parent cell to form a cross-wall, which, when completed, splits sooner or later with separation of the daughter

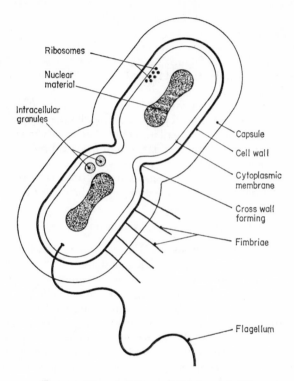

Ribosomes

Nuclear material

Intracellular granules

Capsule

Cell wall

Cytoplasmic membrane

Cross wall forming

Fimbriae

Flagellum

FIG. 1. Diagram of a typical bacterial cell

cells. In many species, e.g. the anthrax bacillus, the daughter cells remain attached and ultimately long strands of many cells are formed since the cell walls remain in continuity; when incomplete splitting of the cell wall occurs the resulting daughter cells remain attached to each other at one point so that a cuneiform appearance is seen on microscopic examination. This is characteristic of diphtheria bacilli.

The cell wall and cytoplasmic membrane play an important part in determining the cell's response to Gram's staining reaction; cells which are Gram-negative, i.e. those that cannot retain the complex of basic dye and iodine in the face of the decolourizing agent, have cell walls which are more porous than Gram-positive cells which resist decolourization.

Other factors are involved in Gram reactivity and these will be noted later; the cytoplasmic membrane usually adheres closely to the cell wall and plays a vital part in absorption of nutrients into and excretion of waste products out of the cell. Similarly, the cytoplasmic membrane contains many enzymes, some of which are associated with respiration whilst others are involved in the production of cell wall material.

Intracellular materials

(1) *Nuclear material*

The 'nucleus' of bacterial cells has the same function as that of nuclei in the cells of plants and animals, i.e. to act as a control centre and to pass on genetic instructions from the cell to its daughter cells. However, the nuclear material of bacteria differs from that of higher animals by not possessing a limiting membrane or a nucleolus and by dividing by simple fission. When bacterial cells are dividing rapidly as in the logarithmic growth phase two, four or more 'nuclei' may be seen in a single cell preceding the formation of cross-walls.

(2) *Ribosomes*

Tens of thousands of these minute granules are present in the cytoplasm of each bacterium; they comprise ribonucleoprotein and a large part of the ribonucleic acid content of the cell is contained in them. Because of their small size they can be seen only with the electron microscope and they are involved in the production of proteins.

(3) *Other intracellular granules*

Volutin granules, consisting essentially of inorganic meta-phosphate may be present in cells and are large enough to be seen with the compound light microscope; these become larger and more abundant when the cell is in a favourable nutrient environment and conversely diminish in number and size and eventually disappear when the cell is in an antagonistic milieu. They would thus appear to function as reserves of food stuffs; the demonstration of volutin granules helps to differentiate certain members of the genus *Coryne-bacterium* from each other.

Lipid granules and others composed of sulphur and gly-cogen may also be demonstrated and their function is prob-ably similar to that of volutin granules; however, they have no significance in identifying species pathogenic to man.

Extracellular structures

(4) *Capsules and loose slime*

Capsules can be formed by certain pathogenic and sap-rophytic bacteria and are circumscribed gelatinous layers outside the cell wall; in pathogenic capsulate species the capsule is most abundant when the bacteria are growing *in vivo* but the production of capsular material is reduced and ultimately ceases on *in vitro* cultivation. Although capsula-tion is a feature of saprophytic as well as pathogenic species there is no doubt that in the latter a strain is more virulent when capsulate; this increased virulence is associated with the ability of such strains to avoid or tolerate phagocytosis by body defence cells. Capsular material consists mainly of water with a very low proportion of solids but the latter, usually complex polysaccharides, are highly specific both

chemically and serologically, so that type specific antisera
can be obtained and used for type differentiation within a
species which is otherwise identical. Reactions between capsular antigens and their specific antibodies are thus employed for the identification of strains of pneumococci and
members of the genus *Klebsiella* for epidemiological purposes.

Capsules are rarely seen in preparations stained by the
ordinary techniques and are most readily demonstrated in
films made with India ink when the bacteria appear as
faint, grey areas within the clear capsules against which
abut the darker ink particles; in such films loose slime can
be noted as irregular masses of varying size extending from
the bacterial bodies (Plate 1, facing p. 56). Loose slime
is formed by many capsulate and some non-capsulate
species and in the former the slime is often very similar
chemically and antigenically to the capsular material;
although loose slime is produced when organisms are growing in fluid media it disperses and cannot be seen, but when
the same organism is cultivated on a suitable solid medium
the secreted slime remains in association with the bacterial
cells and the resulting colony acquires a mucoid appearance
and consistency.

(2) *Flagella* (Fig. 2)

These are the organs of locomotion in all motile bacteria
except spirochaetes; a flagellum is a thin filament twisted
spirally and is usually much longer than the cell which possesses it. Flagella are slender, usually 0.02μ in width, so that
they cannot be seen under the light microscope unless their
width is artificially increased by the deposition of stain on
their surface. In any one species the distribution of flagella
is constant. Some motile bacteria are *monotrichous*, i.e.
they possess only one flagellum per cell which is attached at
one or other pole, e.g. the cholera vibrio, other bacteria
which have a single flagellum at both poles are described

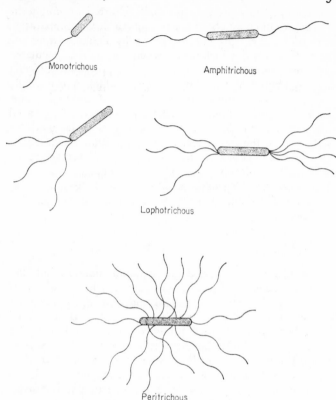

Fig. 2.   Bacterial flagella.

as *amphitrichous*. A *lophotrichous* distribution of flagella is one where two or more flagella extrude from one or both poles of the cell and a *peritrichous* flagellar distribution, i.e. one in which numerous flagella are seen originating from the sides of the cell, is common in several pathogenic species.

Although flagella can be seen with the light microscope in specially stained preparations and also with the electron

B

microscope, their presence is usually inferred *for diagnostic purposes* either by observing motility in a wet preparation viewed by the light microscope or by demonstrating the spreading growth which occurs when the strain is inoculated into semi-solid agar (Plate 2, between pp. 56–7). Flagella consist mainly of flagellin, a protein which is chemically allied to myosin, the contractile protein of muscle in man; as with capsules, highly specific antisera can be prepared against flagellar antigens and this can then be used for determining the serological type of flagellum which a particular strain possesses—this is the basis of serotyping of salmonellae. Motility may allow an organism to escape from unfavourable ecological situations and alternatively to gain access to areas which are beneficial for nourishment and respiration.

(3) *Fimbriae*

These filamentous appendages are approximately half the width of flagella but have no association with motility and occur in pathogenic, commensal and saprophytic bacterial species. They can be viewed only by the electron microscope but frequently their presence can be inferred by the direct haemagglutinating activity of species which possess them. We know very little of their function but since they occur in nonpathogenic as well as pathogenic species it seems unlikely that they have any association with disease production; since fimbriate bacilli adhere to organic particles of various kinds as well as to red blood cells they may benefit the bacterial cell by holding it in contact with a rich source of energy material.

Bacterial fimbriae are a source of confusion in diagnostic serological tests. For example, fimbrial antibodies occur in virtually all individuals and serum from a healthy individual may react non-specifically, often to high titres, with stock diagnostic suspensions of salmonellae if these are fimbriate. Similarly, with certain *Shigella flexneri* isolates type identification may be impossible if the isolate is fimbriate and the diagnostic serum contains fimbrial antibodies because all somatic serotypes of *Sh. flexneri*, if fimbriate, possess identical fimbrial antigens.

Bacterial Endospores

These are usually called spores and are extremely resistant to adverse environmental circumstances which are lethal for the vegetative cells which produce them. The resistance of spores—many will survive boiling in water for several hours—is due to a combination of factors such as the hard spore-case, the low content of free water and extremely low metabolic and enzymic activity.

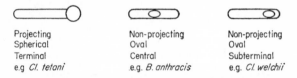

Projecting	Non-projecting	Non-projecting
Spherical	Oval	Oval
Terminal	Central	Subterminal
e.g *Cl. tetani*	.e.g. *B. anthracis*	e.g. *Cl. welchii*

Fig. 3. Size, shape and position of bacterial spores.

Since only one spore is formed by a vegetative cell and on germination of the spore only a single vegetative cell emerges, bacterial spores have no reproductive significance.

Spore formation is restricted to three genera, *Bacillus*, *Clostridium* and *Sporosarcina*, and even then is dependent on environmental circumstances; the fact that the size, shape and position of the spore relative to the vegetative cell are constant in any one species allows the bacteriologist to give a fairly accurate but tentative identification of the species (Fig. 3).

Spores appear as unstained areas in preparations treated by Gram's technique but along with the 'clubs' of *Actinomyces* and members of the genus *Mycobacterium* they share the characteristic of *acid-fastness*, i.e. once they have accepted carbol-fuchsin stain by the Ziehl-Neelsen method they are decolourized only with difficulty, when the preparation is exposed to strong mineral acid; however, spores are the least acid-fast of the structures mentioned since they withstand challenge only by 0.5% H_2SO_4 whereas actinomyces 'clubs' retain carbol-fuchsin when faced with 1% H_2SO_4. Leprosy bacilli are more tenacious and the strength of acid used here is 5% and tubercle bacilli retain the red carbol-fuchsin stain when we attempt decolourization with a 20% solution of sulphuric acid.

Spores, because they can lie dormant for decades, play an important part in the epidemiology of certain human diseases such as anthrax, tetanus and clostridial myonecrosis (gas-gangrene).

STAINING METHODS

Making and fixing a film

Before staining reagents are applied it is essential to spread a film of the specimen on to a glass slide and, after drying, the material must be fixed to the slide. Before spreading the material, e.g. sputum or a suspension of bacterial growth in sterile physiological saline, the bacteriological loop must be sterilized by heating to red-heat in a bunsen flame. The loop is allowed to cool before collecting the specimen with it; the material is then spread on to a 3 in × 1 in glass slide which must be perfectly clean and in particular free from grease otherwise the film will be difficult to spread. It is advisable to keep the film free of the edges of the slide, otherwise the bench and/or the operator's fingers may be contaminated. The inoculating loop is then resterilized as before and the film allowed to dry in air; it may be held high over the bunsen flame to expedite drying.

The dried material is then fixed on to the slide by passing the latter three times slowly through the bunsen flame. By this procedure not only is the film firmly fixed on to the slide but the bacteria present are killed and preserved. It is wise to remember, however, that the cells have been assaulted by drying and heating and their size is much smaller than in the living state; in any case after application of the usual staining procedures the stained material represents only the protoplast since the cell wall is not stained by the methods normally used.

There are many staining methods available for characterizing certain morphological features of the bacterial cell for diagnostic purposes but only two of these are regularly and frequently in use.

Gram's staining method

Gram (1853–1938) described a technique (1884) which allows the very broad classification of bacterial cells into those which react positively, i.e. by retaining the primary dye complex in the face of attempted decolourization, and Gram-negative species from which the primary dye complex is rapidly removed by the decolourizing agent and the cells counterstained with a tinctorially contrasting dye. There are numerous modifications of Gram's staining technique but basically the method depends on the application of a para-rosaniline dye, such as methyl-violet, to a film, followed by washing off the dye with iodine solution which is then allowed to react and 'fix' the dye-stuff so that it cannot be easily removed when decolourization is attempted with acetone or alcohol. Organisms which resist decolourization are Gram-positive and appear as dark purple bodies when viewed with the light microscope; other species are decolourized and are made visible by employing a contrasting counterstain, e.g. basic fuchsin, and these Gram-negative forms are stained pink.

Reference has already been made to the part played by the cell wall and cytoplasmic membrane in Gram's staining reaction but the mechanism of the reaction is not yet completely understood; it may be that the lower pH of the cytoplasm of Gram-positive cells participates in their tenacity for the primary dye complex but whatever the mechanism is, it depends on the physical integrity of the cell and the presence or absence of magnesium ribonucleate in the cell. Cells which normally react positively with Gram's method will become Gram-negative if their magnesium ribonucleate is removed by treatment with ribonuclease; in any Gram-stained preparation of a pure culture of a Gram-positive species there will always be a varying proportion of cells which are Gram-negative and these are cells which have died before the film was made and have thus lost the power of retaining the primary dye complex. The importance of Gram's staining method will become evident in the chapter dealing with classification of bacteria.

Ziehl-Neelsen's staining method

Reference has already been made to the acid-fast nature of certain bacterial materials and the method used to demonstrate acid-fastness was originally described by Ehrlich (1882). Mycobacteria, actinomyces 'clubs' and bacterial endospores are relatively impermeable to ordinary dye-stuffs but in the Ziehl-Neelsen method basic fuchsin will penetrate such structures provided that phenol is present and the film is heated; once stained red by this method tubercle bacilli and other acid-fast structures will resist the decolourizing action of strong acids for a time greater than that of other non-acid fast material in the film.

Methylene blue is commonly used to counterstain tissue cells and bacteria which are decolourized by the acid. The differentiation of saprophytic from pathogenic mycobacteria can be made at a microscopic level since pathogenic species are also alcohol-fast and attempted decolourization with 95% ethanol following the application of 20% H_2SO_4 allows such species to retain the red carbol-fuchsin dye, whereas alcohol-decolourization will leach out the carbol-fuchsin from saprophytic species. This has practical importance when specimens of urine are being examined microscopically in cases of suspect renal tuberculosis since commensal acid- (but not alcohol-) fast smegma bacilli may be present; acid-fastness is closely associated with the presence in the intact cell of lipids which, along with fatty acids, abound in acid-fast bacteria since the staining property is lost when the lipids are removed from the cell.

CHAPTER 3
Bacterial Physiology

Whereas the academic study of the life processes of bacteria is now properly the interest of the fundamental bacteriologist, the medical bacteriologist's interests in the physiology of bacteria are very much of an applied nature since he is concerned primarily with producing optimal conditions for the rapid isolation and identification of pathogenic species. Hence in this chapter we shall pay attention mainly to the applied aspects of the subject.

In their life processes bacteria have many points of similarity with higher organisms, including man. They therefore require nourishment and must respire and reproduce. In the same way they respond to environmental factors such as heat, light, sound and other potential hazards.

In the previous chapter we noted the complex biochemical nature of the bacterial cell and when it is remembered that under optimal conditions most bacteria reproduce themselves every 15–30 minutes it is obvious that the cells require almost limitless sources of food materials and energy.

Nutrition

In general terms the chemical composition of all bacteria is very similar, yet there is wide variation in the basic nutrients required by different species; these differences in requirement for cultivation *in vitro* reflect the natural environmental adaptations of the different species and thus their varying abilities to synthesis materials.

Certain non-parasitic bacteria, like plants, are able to rely on CO_2 as a main carbon source for growth and may obtain their energy for synthetic processes from sunlight (the photo-synthetic autotrophs) or by the oxidation of inorganic material (the chemo-synthetic autotrophs). However, most bacteria, including all species which are parasitic on man, are unable to utilize such elementary sources of carbon and energy and must be supplied with organic nutrients; such species are called heterotrophs.

There is wide variation in the range of organic compounds which heterotrophic bacteria can use and in general terms it can be stated that the more specific or exacting is a species in its nutritional requirements the more parasitic it has become and the ultimate level of parasitism is of course exhibited by viruses which are entirely dependent on other living cells to provide their food stuffs in prefabricated form.

In addition to carbon, hydrogen, oxygen and nitrogen (which are the main elements necessary for the growth of bacteria) other materials are required in smaller amounts, e.g. sulphur, sodium, iron, etc; furthermore some essential metabolites are required in almost infinitesimal quantities since they function essentially as catalysts. In many instances these essential metabolites are the same as the vitamins necessary for mammalian nutrition and are frequently referred to as 'bacterial vitamins'.

Respiration

A few bacteria, e.g. the tubercle bacillus, are described as *strict* or *obligate aerobes* since they grow only in the presence of free oxygen or air and alternatively members of certain genera, e.g. those of the genus *Clostridium*, cannot grow if even traces of air or free oxygen are present in their environment and they are spoken of as *strict* or *obligate anaerobes*; the vast majority of bacteria, however, can grow whether or not their atmosphere contains oxygen and are referred to as *facultative anaerobes*. Lastly several organisms are *microaerophilic*, e.g. *Brucella abortus*, and grow best when only a trace of oxygen is present and many such species prefer that the CO_2 concentration of their atmosphere should be increased to 5–10%. All bacteria, of course, require CO_2 for growth but the amount present in the normal atmosphere is usually sufficient.

Temperature

Almost all bacteria which are pathogenic to man have an *optimum temperature* for growth of 37°C and this reflects their adaptation to a particular host. However the range of temperature over which growth occurs varies widely; the temperature range for growth may be from a very low minimum, e.g. 5°C to a high maximum, 43°C as in the case of *Pseudomonas pyocyanea* or over a very restricted range, e.g. 30°–39°C as for gonococci. In general, the narrower the temperature range the higher is the degree of parasitism and the more nutritionally exacting is the species.

Bacteria which grow best within a range of 25° and 40°C are termed *mesophilic* and include all of the species parasitic on man and warm-blooded animals as well as many species saprophytic in soil and water. *Psychrophilic* bacteria, i.e. those which grow best at temperatures below 20°C, are non-pathogenic for man but exist in soil and water; certain psychrophilic species can grow even at temperatures below 0°C and therefore have an opportunity to grow on and spoil foodstuffs held in cold storage. *Thermophilic* species can flourish at temperatures of 55°–80°C and like psychrophilic bacteria are non-pathogenic to man.

When a bacterium is maintained at a temperature lower than the minimum at which growth occurs it usually survives and indeed almost all species, including those pathogenic to human beings, can survive refrigeration; this extraordinary biological characteristic is invaluable to the bacteriologist since cultures can be lyophilized, i.e. freeze-dried and maintained in this state for very long periods of time without sub-cultivation.

In contrast to their ability to survive exposure to low temperatures, bacteria of any species are killed by exposure to temperatures significantly higher than the maximum at which growth occurs and for each species a thermal death point (TDP) can be determined. The TDP is defined as the lowest temperature above the maximum at which growth occurs at which a species is killed in a given period of time, e.g. ten minutes. For non-sporing mesophilic species the TDP varies from 50°–56°C on exposure to moist heat.

Bacteria which form spores have a TDP of 100°–120°C and the most resistant spores are those of *Bacillus stearothermophilus* which must be exposed to 121°C for 20 min before they are destroyed.

Hydrogen ion concentration

The vast majority of parasitic bacteria grow best at a slightly alkaline pH, i.e. 7.2–7.6, although most species are tolerant of a wider range; some bacteria, e.g. Döderlein's bacillus are *acidophilic* and can grow at a pH of 4. Other species, notably the cholera vibrios are intolerant of an acid pH and prefer a highly alkaline environment; in the case of *V. cholerae*, media for *in vitro* cultivation have an initial pH of 8.5.

Moisture

There is a wide variation in the ability of bacteria to survive drying; bacterial spores are extremely resistant to desiccation and can survive for many years in a moisture-free environment. Vegetative cells are much less resistant to drying and, for example, gonococci die within an hour or two of leaving their human host even if they are in a favourable pabulum unless the latter has a high moisture content; other non-sporing species can survive for weeks or even months in inanimate surroundings provided they are not subjected to hazards other than drying, e.g. tubercle bacilli can be recovered from floors, bedclothes, etc, for two or three months after they have been expectorated by a tuberculous patient.

Radiations

All pathogenic bacteria are susceptible to the lethal effects of ultraviolet rays whether naturally exposed to direct sunlight or to an artificial source such as a mercury vapour lamp. Of course, the saprophytic photo-synthetic autotrophs require sunlight for their life processes and here again we may note the very wide spectrum of the influence which environmental factors play in the microbial world.

Many other kinds of radiation are lethal to bacteria and gamma-radiation from a Cobalt 60 source is now used to sterilize heat-labile materials such as disposable plastic syringes.

The cultivation of bacteria

In attempting to isolate bacteria from pathological or other material the bacteriologist has to cater for the above physiological requirements in various combinations.

The nutritional requirements are met by providing various culture media; the simplest medium is a *broth*, e.g. nutrient broth, which is a mixture of peptone and meat extract in water. *Peptone* is obtained from the digestion of meat with a proteolytic enzyme, e.g. trypsin and the mixture of polypeptides and amino acids thus released serves as a source of nitrogen, carbon and energy. *Meat extract*, i.e. the water-soluble components of meat, provides various mineral salts and bacterial vitamins.

Since organisms growing in such a broth cannot form colonies, and thus we cannot determine whether the culture is pure or consists of a mixture of organisms, the fluid medium can be solidified by adding a small amount of agar and the resulting medium, i.e. nutrient agar, is usually dispensed in sterile Petri dishes.

Media, whether fluid or solid, can be made more nutritive by adding other materials such as blood or serum which allow the growth of more exacting species. Similarly, by incorporating various chemicals media can be made more selective for a given species, e.g. the addition of potassium tellurite to a blood agar plate prevents the growth of many bacteria whilst allowing diphtheria bacilli to flourish.

In preparing culture media the final pH is adjusted to suit the requirements of the majority of species, i.e. a pH of 7.2, and for particular species, as mentioned above, the hydrogen ion concentration can be altered to assist the growth of organisms which may prefer an environment beyond the normal, near-neutral range.

When inoculating solid media in Petri dishes it is essential to plate out the material as shown in Plate 3 (between pp. 56–7) so that the resulting colonies will be well separated and if more than one species is isolated then each can be obtained in pure culture for further study.

The inoculated media are then placed in an incubator maintained usually at a temperature of 37°C—the optimal temperature for almost all bacteria pathogenic to men; incubators are light-proof so that any possible harmful effects of ultraviolet radiation are eliminated. The normal atmosphere in the incubator will suffice for the majority of bacteria but when we are attempting to isolate strictly anaerobic species an oxygen-free environment must be provided; this is most readily obtained by enclosing the inoculated media in an anaerobic jar. After sealing the jar a high vacuum is drawn, e.g. 30 cm of mercury and hydrogen gas is then introduced and any residual atmospheric oxygen is eliminated either by passing a low voltage electricity supply through a protected electric coil or in more recent types of anaerobic jar the slow union of hydrogen and residual oxygen is effected by a cold catalyst comprising basically palladium chloride.

In either instance it is essential to include in each jar an indicator of anaerobiosis; essentially such an indicator comprises methylene blue which can be altered to its colourless state by boiling immediately before it is placed in the anaerobic jar. If on opening the jar after incubation there is any return of the natural blue colour then anaerobic conditions have not been effected.

CHAPTER 4
Aggressive Mechanisms of Bacteria

Here we consider bacterial factors which are involved in allowing organisms to attempt to establish themselves in or on host tissues; necessarily these have to be considered separately and in isolation from the host defence mechanisms which are dealt with in the next chapter. However, it is important to remember that the interaction of bacterium and host is complex and, also, that there are still large areas of ignorance not only of the interaction but also of the mechanisms of the two interacting forces.

In contrast with *saprophytic* bacteria which live freely in nature on decaying organic matter, in the soil or in water, *parasitic* bacteria live in or on a living host. Parasitic bacteria may lead a *commensal* existence with their host and occasionally with mutual benefits or a parasite may be *pathogenic*, i.e. produce disease in the host.

Some commensal species can cause disease due often to an alteration in the host's tissues, e.g. *Strept. viridans*, a normal inhabitant of the mouth and throat, is the commonest cause of subacute bacterial endocarditis, but this disease occurs usually only if the host has a predisposing heart lesion and is in a poor state of dental hygiene; similarly pathogenic species may inhabit host tissues without causing any disorder, e.g. coagulase-positive staphylococci are carried in the anterior nares by a significant number of the population, particularly hospital personnel, without any lesions resulting.

It will be obvious, therefore, that the distinction between commensal and pathogenic bacteria (and other microorganisms) is by no means clear cut and that the detection of a bacterial species in host tissues or exudates is not necessarily equated with disease.

In order to rationalize this situation and be able to state, with reasonable assurance, that a particular disease is caused by a given bacterium a series of guide-lines, known as *Koch's Postulates* are often invoked.

These postulates are that before an organism can be accepted as the cause of a disease (1) it should be found in all cases of the disease and its distribution in the host's body should be in accordance with the lesions observed, (2) the organism should be isolated *in vitro* in pure culture, (3) the pure culture should reproduce the disease when introduced to a suitable experimental animal and (4) it should in turn be isolated in pure culture from that animal.

Although there are some infections, e.g. leprosy, in which only the first of these criteria of pathogenicity can be fulfilled, with improving techniques we can usually satisfy the first two postulates; however, since many parasitic bacteria show a high degree of host specificity, e.g. the typhoid bacillus, it is often impossible to reproduce the disease in an experimental animal and hence the last two postulates are frequently not fulfilled. One could, of course, do so even with a bacterium of high host specificity for human beings if man came into the category of 'a suitable experimental animal'.

However, there are two pieces of very strong circumstantial evidence which frequently can replace our inability to prove the last two of Koch's postulates; the demonstration of a significant increase in the host's serum antibody level to a particular bacterium is frequently invoked as a fifth postulate and as a corollary a host recently recovered from infection is highly resistant to fresh challenge with the strain causing the infection. The second piece of circumstantial evidence concerns the epidemic spread of a disease when although only the first two postulates may be capable of fulfilment the constant isolation of a given bacterium from each case and the similarity of each host's signs and symptoms leaves little doubt as to the cause-and-effect relationship between bacterium and disease.

FACTORS ASSOCIATED WITH PATHOGENICITY

It must be emphasized immediately that in many instances we still await the discovery or definition of what features of the bacterial cell are associated with pathogenicity.

Surface layers

There are several examples of the important part played by *bacterial capsules* in determining whether or not a particular species is pathogenic; pneumococci freshly isolated from pathological material are heavily capsulated and

therefore relatively resistant to phagocytosis and when injected into a non-immune mouse cause septicaemia and
rapid death. If such pneumococci are subcultured in the
laboratory non-capsulate mutants can be derived which are
avirulent for mice, hence bacterial capsules, by protecting
the bacteria from phagocytes, are one of the factors involved in pathogenicity. Similar evidence to that for pneumococci has demonstrated the importance of capsules in
determining the pathogenicity of anthrax bacilli, group-A
streptococci, Friedländer's bacillus and several other bacteria. No matter how important capsulation may be in
determining the pathogenicity of certain species it cannot
be *equated* with pathogenicity since some entirely saprophytic bacteria produce capsules.

In the case of typhoid bacilli, pathogenicity is associated
with the presence of a non-capsular surface layer of Vi
(virulence) antigen and strains which lack this antigen are
avirulent; the mechanism involved is not known.

Specialized surface layers, e.g. capsules and microcapsules,
can, however, lose their protective ability when the organism
in question meets specific antibodies and complement; similarly somatic antigens can have their effect nullified.

Toxins

Many, if not all, bacteria produce substances which cause
damage to host tissues; some species, e.g. diphtheria and
tetanus bacilli which are only weakly invasive and usually
cannot spread beyond the initial locus of infection produce
dramatic clinical effects in the host by virtue of the *exotoxins* which they produce; this group of toxins diffuses
from living bacteria into the environment and, *in vitro*, can
be separated by centrifugation of the fluid medium in which
the bacteria have been grown. Exotoxins thus harvested
from the supernatant of a broth culture, have so far been
shown to be simple proteins, the majority are unstable
in that they can be rendered non-toxic, i.e. toxoid pre-

parations can be derived (e.g. by treatment with formalin) which retain antigenicity and are useful as active immunizing agents; the antitoxic antibodies thus produced are
usually highly protective to the host. Many exotoxins are
extremely potent with a lethal dose for men of 0.0005 mg or
less and often have a highly specialized action, e.g. teta-
nospasmin, the lethal exotoxin produced by the tetanus
bacillus, acts on the motor nerve cells. Such exotoxins are
produced mainly by Gram-positive bacteria.

By contrast *endotoxins* are released from bacteria only
when the cells die and are not found in significant amounts
in bacteria-free supernatants unless the organisms have disintegrated either naturally or under some artificial influence. Structurally endotoxins are complex and their
toxic action is similar regardless of the species which produce them, they are more stable than exotoxins and immunization with endotoxins produces *antibacterial* antibodies which unlike antitoxic antibodies cannot nullify or
protect against the toxic effects of endotoxins. Both Gram-
positive and Gram-negative species produce endotoxins but
those of Gram-negative species are most potent.

Toxins are often named according to the effects they
produce, e.g. haemolysin (lysis of red blood cells), necro-
toxin (tissue necrosis) etc.

The important role of exotoxin in producing disease
has already been mentioned in the case of diphtheria
bacilli; even more significant is the fact that certain syndromes, e.g. botulism, can occur without the causal organism itself being present in the host's tissues; here, botulinus
toxin affecting the central nervous system, may be present
in foodstuffs ingested after the toxin-producing strain has
been destroyed.

Other bacterial products

There are several other bacterial products which, *per se*, are
not toxic, but by acting on host tissues may facilitate the

c

spread of infection; fibrinolysins are produced by many
species, e.g. group-A streptococci and may allow more
speedy invasion of tissues with resulting cellulitis in comparison with more localized surface infections which usually
result with species which have little or no fibrinolytic
activity.

Many bacteria produce diffusion factors which are probably one and the same namely, hyaluronidase; this enzyme,
by hydrolizing the hyaluronic acid which is an intercellular
tissue substance, can promote the spread of bacteria through
tissues. As with most 'pathogenicity factors' there are
several highly invasive bacterial species which do not produce hyaluronidase, thus there is not a complete correlation
between hyaluronidase production and invasiveness.

CHAPTER 5
Host Defence Mechanisms

Immunity may be defined as the ability or power of the
host to resist infection by bacteria and/or the harmful
effects of their toxins. Thus immunity may be natural (in-
nate) or acquired, and may be specific for a given bacterial
species or non-specific against a wide range of bacteria and
other micro-organisms.

NON-SPECIFIC NATURAL IMMUNITY

In the healthy individual pathogenic bacteria can gain
access only through the skin or one of several mucous membranes and these are in the front line of the normal non-
specific host defence mechanisms.

Skin

Skin offers a physical and chemical barrier to bacteria. The *physical barrier* comprises firstly the dead, outer keratinous layer which offers protection against bacteria to the underlying, living epithelial cells and also the thickness of the skin through which no bacteria can penetrate unaided, with the possible exception of leptospirae. This physical barrier has, however, innumerable weak points in the form of hair follicles and gland openings through which many bacterial species may penetrate deeply into the dermis.

The *chemical barrier* consists of long-chain fatty acids which maintain much of our skin surface at a pH of 6 or less and creates an environment inimical to many bacteria but there are gaps in the acid barrier, notably in the axillae and groins and these areas, which also have a dense population of hair follicles, may be more readily colonized than the general skin surface.

Mucous membranes

Because of their finer structure these may be thought to be more liable to attack by bacteria than is the skin but the thickness of mucous membranes is enhanced by the mucous secretions which trap bacteria which are then swept along and excreted from the particular tract.

Mucous membranes also possess additional protective mechanisms and those of the various tracts are considered briefly below.

Respiratory tract

The vast majority of organisms breathed in through the nose do not pass beyond the anterior nares and the few which do so are trapped by the nasal mucosa and wafted towards the pharynx by the action of the ciliated epithelium. Very few bacteria persist in these areas and in addition to mechanical removal it is probable that some antibacterial chemical action is involved.

Bacteria which gain access to the mouth are rapidly swept backwards by suction from the tongue, palate, etc. and join those which escaped the filter in the anterior nares; thus many organisms introduced into the upper respiratory tract will be swallowed and subjected to the defences offered by the stomach.

In health probably less than 5% of bacteria which are inhaled pass beyond the upper respiratory tract and those which pass beyond the larynx are trapped in the mucus and then removed upwards by ciliary action and expectorated; the few which reach the lung alveoli are normally destroyed by phagocytic cells.

Intestinal tract

Human saliva has some antibacterial activity but organisms swallowed in foodstuffs will frequently meet their first challenge in the stomach when they encounter the highly acidic gastric juices which are lethal to the vast majority of bacteria. Obviously, if food is chewed thoroughly instead of being swallowed as a bolus, bacteria in it will be more readily destroyed in the acid environment; lactobacilli are commensal in the intestine and would appear to have a defensive rôle against infection since if they are eliminated, e.g. by broad-spectrum antibiotics given pre-operatively to 'sterilize' the bowel, the consequences may be serious if they are replaced by staphylococci or yeasts which can, under these circumstances, give rise to severe infection.

Genito-urinary tract

With the exception of a small number of staphylococci and other essentially commensal species which can be found around the external meatus, the urethra is normally sterile and frequent flushing during micturition is probably the most important factor in maintaining sterility; the acid pH (5–6) of urine in healthy individuals is also a protection against bacterial infection.

The vagina, particularly from puberty until the menopause, has a most efficient defence mechanism provided by a commensal bacillus, Döderlein's bacillus, which produces a highly acid vaginal secretion by fermenting the glycogen of the vaginal epithelium.

The conjunctivae

The shuttering action of the eyelids combined with the flushing by tears acts as a very efficient mechanical barrier; in addition tears have a very high lyzozyme content.

Bacteria which manage to breach any of the surface defence mechanisms summarized above have to run the gauntlet of other equally formidable barriers if they are to establish themselves in the host tissue.

Phagocytosis

Phagocytosis, i.e. the ingestion of bacteria by certain body defence cells, is a general, non-specific cellular defence mechanism. Bacteria may be ingested by *wandering* phagocytes, i.e. polymorphonuclear leucocytes and large mononuclear leucocytes, and if susceptible are usually destroyed by enzymes produced by the leucocytes; phagocytosis is also an activity of the *fixed* macrophages of the reticulo-entothelial system in the spleen, liver and bone marrow after bacteria which have escaped the wandering phagocytes leave the blood stream and spread throughout the tissues. To do so, of course, the bacteria must have successfully avoided ingestion by reticulo-endothelial cells in the lymph-nodes draining the area of primary invasion.

A detailed discussion of the mechanism of phagocytosis is the privilege of the pathologist but it should be noted that not only are capsulated organisms resistant to the process but certain other bacteria, e.g. tubercle bacilli and salmonellae, although ingested by defence cells, can multiply within them and destroy the phagocyte.

In addition to the non-specific *cellular* defence offered by phagocytes of one kind or another, non-specific *humoral* mechanisms have been identified. As an example of the latter we take *properdin* which has a wide activity not only against bacteria but against certain viruses, yeasts and protozoa and is present in normal serum; thus it is accepted as part of the non-specific resistance which the human host possesses against infection. Since properdin is a high molecular weight protein or group of proteins and can cause lysis of the bacterial cell only when serum complement is present it may, in fact, prove to be simply a mixture of several specific antibodies.

Basic proteins

A host of basic proteins have been described which are released from tissue cells destroyed during infection and have antibacterial activity without the aid of serum complement. These are active mainly against Gram-positive cells. Another basic protein present in *healthy* tissues is lyzozyme and although its mucolytic activity is most readily displayed against certain saprophytic bacteria it can attack the cell walls of certain pathogens, e.g. anthrax bacillus.

SPECIFIC ACQUIRED IMMUNITY

In contrast to the non-specific resistance factors so far discussed, all of which are naturally occurring phenomena, an individual may acquire resistance which is highly specific since it gives protection against one particular bacterial species or exotoxin.

This specific acquired immunity is dependent on the presence of specific antibodies in the individual's blood serum and tissues; antibodies are serum proteins, usually γ globulins, so modified that they react specifically with the particular antigens which stimulated the host's reticulo-endothelial cells to form them.

When an individual recovers from a natural infection, antibodies specific for the infecting agent are usually demonstrable and such acquisition of immunity is termed *natural, active* immunity because the host's tissues have actively produced the antibodies. Alternatively, *natural* immunity may be *passively* acquired by the foetus by placental transfer of antibodies from the mother.

Immunity may be acquired *artificially* and again this may be *active* when antigenic material is administered to the individual whose tissues then manufacture the antibody, or it may be *passive*, i.e. when antibodies produced in another host are injected.

Active immunity, whether resulting from infection or from administration of antigenic preparations develops more slowly than *passive* immunity, which is acquired very rapidly. When endowed artificially the speed with which *passive* immunity develops depends primarily on the route of injection of the antiserum, being virtually instantaneous following intravenous administration and requiring only a few hours when administration is intramuscular or subcutaneous. However, active immunity, once established, lasts at least months and often many years, whereas passive immunity is of short duration. When passive immunity is *naturally acquired* by the foetus from its mother, antibodies can still be detected in the baby for some months (4–6) after birth. When, however, passive immunity is *artificially acquired* by injection of antibodies derived from animals or other human beings, the immune state is very brief and rarely lasts more than one month since the antibodies are even more foreign to the recipient than maternal antibodies and are very quickly destroyed by the individual.

Further discussion of various aspects of specific acquired immunity takes place in the next chapter.

CHAPTER 6
Prophylactic Immunization

A brief account of the host's defence mechanisms has been given in Chapter 5 and in several of the chapters dealing with the various bacterial genera reference will be made to the protective value of immunization procedures; in this present chapter we offer a résumé of certain of the practical aspects of prophylactic immunization.

HISTORICAL INTRODUCTION

Perhaps the first recorded attempt to artificially immunize individuals against an infection was that of Francis Home in Edinburgh in 1758; Home attempted to produce a modified form of measles in healthy children by inserting into skin incisions cotton threads which had been soaked in the blood of patients in the florid stage of the natural disease. Forty years later Edward Jenner published his proof of the protective influence of cowpox against an attack of smallpox; the fact that people who had suffered from cowpox rarely, if ever, contracted smallpox had been noted by members of farming communities long before Jenner confirmed the fact by scientific experiment but the excellence of his work rightly allows him to be regarded as the father of modern methods of artificial immunization.

Earlier attempts to protect against smallpox had been practised in many countries and in 1721 Lady Mary Wortley Montagu introduced the practice of variolation from Turkey; variolation, i.e. the introduction into the skin of a healthy person of material taken from the vesicles of a case of smallpox, had certainly been practised in certain parts of Britain before Lady Montagu popularized its use. Obviously variolation and Home's method of protecting against measles offered no control of the number of virus particles introduced into the healthy person and as the virus particles were fully virulent, modified attacks of these diseases did not always result and severe and even fatal infections were not uncommon in the variolated subject. Home's method of protecting children against measles never attracted attention but variolation was practised until it was made illegal in 1840; by that time it had been accepted that Jennerian vaccination was much more successful than variolation and was very much less dangerous, not only to the individual but also to other people in the community since a variolated person sometimes acted as a source of epidemic smallpox.

With the exception of vaccination against smallpox the agents which we presently use for active immunization against infectious diseases were developed only after the discovery of the causal microorganisms and the invention of methods which allowed the development of safe, killed or attenuated vaccines; similarly the introduction of the hypodermic syringe by Alexander Wood in 1853 preceded the widespread use of vaccines, the majority of which require parenteral administration.

POTENTIAL HAZARDS OF IMMUNIZATION PROCEDURES

As with all discoveries that of the hypodermic syringe has disadvantages; apart from the prospect of injecting material into a blood vessel or causing injury to a nerve whilst administering antigenic substances there is the risk of causing infection unless a *separate, sterile syringe and needle* is used for each person being immunized. Local sepsis at the injection site is probably more common though less serious than the other iatrogenic infection which can occur namely, serum hepatitis; serum hepatitis is transmitted from a carrier or case of that infection by a syringe or needle which is used to inject a healthy person without previously being cleaned and sterilized.

The amount of blood required to transmit serum hepatitis is minute, not more than 0.01 ml, so that transmission can also occur via stilletes used for finger pricking to obtain blood for haematological investigation.

In many circumstances it is more convenient to use sterile, disposable plastic syringes for immunization purposes rather than to rely on a supply of freshly sterilized all-glass syringes; similarly when large numbers of people are to be immunized wider use might be made of needleless, high-pressure jet injectors.

Many calamities have occurred in the past, e.g. because of inadequate control of the production of vaccines whereby live microorganisms have survived in an allegedly killed vaccine; in almost all countries however very strict control of the potency and safety of vaccines is now undertaken.

Nevertheless virtually all immunizing agents have potential complications and some of these are noted below; the risk of such complications is statistically small and acceptable when the disease against which protection is sought carries a high case fatality rate and/or is endemic in a community. When such endemic diseases are eliminated from a community (often as a result of active immunization) then the population becomes less willing to run the risk of complications resulting from continued efforts at artificial protection.

Complications of smallpox vaccination

Probably the most common complication in this instance is local sepsis resulting from the implantation of pathogenic bacteria at the time of vaccination or alternatively such bacteria may gain entry after the vaccinial lesion has appeared; the risk of local sepsis from simultaneous implantation of pathogenic bacteria is lessened if the multiple pressure method of vaccination is employed instead of the older scarification technique.

More serious and fatal complications are firstly, generalized vaccinia which occurs most commonly after primary vaccination and particularly if this is delayed until adolescence or adulthood; however generalized vaccinia has its highest fatality in infants under 1 year of age and this is one of the main reasons why, in countries where smallpox is no longer endemic, vaccination against the disease should be delayed until after the first birthday but should be performed before school entry. Also the incidence of generalized vaccinia is lowest in the 1–4 year age group.

A second serious complication of vaccination against smallpox is post-vaccinial encephalitis; it is probable that this complication is not caused directly by the vaccinia virus but is due to another neurotropic virus stimulated into activity by the vaccinial lesion. The incidence and fatality rates of post-vaccinial encephalitis are also minimal in the 1–4 year age group. Such complications are much less commonly encountered following revaccination of the individual regardless of the age of the person; a less important reason for delaying primary smallpox vaccination until the later pre-school years in our society is that the first year of life is increasingly occupied with other immunization procedures.

Provocation poliomyelitis

In 1950 three separate reports drew attention to the fact that a number of patients with paralytic poliomyelitis had received prophylactic inoculations shortly before they developed poliomyelitis and that their paralysis was either restricted to or concentrated in the limb used for inoculation; although this sequence of events might have been coincidental it was soon shown that the paralysis had in fact been provoked by the immunizing injections.

Procedures other than active immunization are also recognized as capable of provoking poliomyelitis, e.g. intramuscular injection of penicillin, although in such instances provocation is usually only noticed when large numbers of people in a community are almost simultaneously injected as in a yaws eradication campaign. Tonsillectomy and dental extraction have also been associated with provocation poliomyelitis and obviously such operative procedures should be restricted during epidemics of poliomyelitis.

The risk of provocation poliomyelitis can be reduced by ensuring that protection with an orally administered attenuated live vaccine is given early in the child's immunization schedule.

Complications of immunization against whooping cough

The incidence of local sepsis following the injection of pertussis vaccine is not known but perhaps the most common complication of vaccination against whooping cough is febrile convulsions with an incidence of 1.2 per 1000 children immunized with three injections; by contrast febrile convulsions occur in approximately 5% of unimmunized children before their fifth birthday and in association with some infection, frequently of the upper respiratory tract. Thus the risk of febrile convulsions following immunization with pertussis vaccine is much less than such convulsions occurring after naturally acquired infections and fatalities are very rare; nevertheless children with a history of febrile convulsions should normally not be given pertussis vaccine.

A much less common but more serious complication is encephalopathy and although we do not have any accurate measurement of the incidence of post-immunization encephalopathy it must be infinitesimal when one considers the hundreds of millions of doses of pertussis vaccine given in the last thirty or more years; by comparison the recorded and proven cases of associated encephalopathy number only

100 or so. Encephalopathy as a complication of the natural disease has an incidence of at least 0.8% and this risk is very much greater than that following active immunization against whooping cough.

SOME EXAMPLES OF THE VALUE OF PROPHYLACTIC IMMUNIZATION

Perhaps the most spectacular example of the influence of active immunization in recent years is that of the dramatic control of diphtheria following mass immunization of the community. Diphtheria was first made notifiable in Scotland in 1889 and during the present century the annual number of cases notified varied from approximately 8,000 to more than 11,000; mass immunization commenced in Scotland in 1940 and the annual incidence had dropped to less than 4,000 cases by 1946 and to less than 400 three years later. In 1959 only 2 cases of diphtheria were notified and no further cases have occurred; similarly evidence of the value of immunization against this infection may be noted from the differential incidence and mortality figures in those who had been immunized in comparison with non-immunized members of the Scottish community.

	CONFIRMED CASES		DEATHS	
Year	Non-immunized	Immunized	Non-immunized	Immunized*
1942	6,956	1,799	514	3
1946	2,122	1,024	85	6
1949	272	61	14	–
1959	2	–	1	–

* The last death from diphtheria in an immunized person was in 1948 although there were 34 deaths from the disease in non-immunized individuals between 1949 and 1959.

The evidence of the protective influence of immunization with diphtheria toxoid is overwhelming but in other infections the prophylactic value of active immunization may not be so dramatic and in such instances it is necessary to evaluate the immunizing agent by setting up a controlled trial in a community where the infection is endemic.

In such controlled trials a test group of people to be given a particular prophylactic agent is matched with another group, the control group, which is as biologically equivalent as possible with the test group; both groups should have the same age and sex composition and be drawn from similar social backgrounds. In addition both groups should contain families of similar size and the experience of both groups in regard to past infections should be virtually identical.

Prevention of tuberculosis

Bacille Calmette-Guérin (BCG) vaccine was first advocated as a means of protecting against natural tuberculous infection by Calmette and Guérin in 1906 and the vaccine was used for several decades in many countries although conclusive evidence of its efficacy was not obtained. Controlled trials of BCG vaccine were first undertaken when the Medical Research Council in Britain commenced evaluation of the vaccine in 1949; these trials have shown conclusively that in a country such as Britain with highly developed health services BCG vaccination gives at least a 79% reduction in the incidence of tuberculosis in comparison with an unprotected control population living in similar circumstances.

Another vaccine, the Vole vaccine derived from the murine type of tubercle bacillus, was also incorporated in the trials at a later stage and the Vole vaccinated test population showed an 81% reduction in the incidence of tuberculosis as compared with the control non-immunized group.

Both BCG and Vole vaccines appear to be absolutely protective against the more severe forms of tuberculosis such as miliary infection and tuberculous meningitis.

Without wishing to detract from the value of BCG or Vole vaccination it should be noted that measures other than vaccination have contributed significantly to the reduced incidence of tuberculosis, particularly infection derived from bovine sources, i.e. the creation of cattle herds free from tuberculosis combined with the increasing practice of pasteurizing milk for human consumption.

Immunization against the enteric fevers

Before any evaluation of vaccines is made in the human population they are almost invariably assayed for potency etc. by animal experiment; however the results of such tests have on occasion been misleading. For example, in the early 1940's an alcohol-killed and -preserved TAB vaccine (alcoholized vaccine) was found to give a much higher degree of protection against challenge with virulent typhoid bacilli in immunized mice than was obtained in an identical mouse population which had been given the usual heat-killed, phenol-preserved TAB vaccine (phenolized vaccine); this enhanced protection of the alcoholized vaccine was associated with the fact that such a method of preparing the vaccine retained the virulence or Vi antigen and that mice immunized with alcoholized vaccine produced Vi antibodies. Mice receiving the phenolized vaccine did not produce Vi antibodies.

However, when in 1954 field trials of these two types of TAB vaccine were conducted in Yugoslavia it was soon apparent that, regardless of the results of the mouse experiments, the newer alcoholized vaccine had little more protective effect in the human population than had a vaccine prepared from *Shigella flexneri* which was given to the control group in the population.

The group which had received phenolized TAB vaccine

had an attack rate from typhoid fever of 6.1 per 10,000 whereas the attack rates for the groups receiving alcoholized TAB vaccine and the *Sh. flexneri* suspension were 14.1 and 19.2 per 10,000 respectively. Subsequent controlled field trials with the phenolized vaccine and a more recently developed acetone-killed TAB vaccine in Guiana revealed that the latter gave an even better protection than the phenolized product since the average annual typhoid fever attack rate was only 1 per 10,000 of the group receiving acetone-treated TAB vaccine.

In these trials of TAB vaccines there was no correlation between the antibody levels which individuals developed as a result of immunization and their protection from typhoid fever; a proportion of people with very low antibody response escaped infection whereas others with high agglutinating antibody levels succumbed to natural infection with *S. typhi*. This latter finding is not altogether surprising since people recovering from the natural infection, and with high antibody levels, occasionally suffer a relapse of the infection in the late clinical stage or in early convalescence.

Such findings underline the need for evaluation of vaccines by controlled field studies in the human population which allow the assessment of attack rates in the protected and non-immunized groups.

IMMUNIZATION SCHEDULES

There are now at least 19 communicable diseases against which a more or less satisfactory degree of protection can be offered by active immunization; fortunately no community requires that the population should be given protection against all of these infections and indeed in some cases only individuals at special risk require to be protected, e.g. those who may be exceptionally exposed because of their occupation. Such 'special risk' diseases against which one can offer artificial immunity are noted below.

Anthrax	Measles	Tularaemia
Brucellosis	Mumps	
Leptospirosis	Q-fever	

The other infections against which protection may be afforded by artificial active immunization are:

Cholera	Plague	Tuberculosis*
Diphtheria*	Poliomyelitis*	Typhoid fever
Influenza	Smallpox*	Typhus fever
Pertussis*	Tetanus*	Yellow fever

Those infections against which children in Britain still require protection are asterisked but the above list should also remind us that diseases such as cholera, plague and typhus fever rampaged through this country until relatively recently and were driven out by measures such as improved sanitation and safe water supplies; the continuing presence of these same infections in other countries underlines the need for such measures as well as the practice of preventive medicine.

Recommendations regarding which vaccines should be offered to a community and the priority in which vaccines are given can only be made with a knowledge of the morbidity and mortality rates of the several diseases; obviously since whooping cough, although uncommon in the first six months of life, is restricted in its lethality to the first year then pertussis vaccine must be given precedence in Britain. Fortunately this vaccine can be combined with diphtheria toxoid and tetanus toxoid so that the young infant can be simultaneously protected against three diseases.

The following immunization schedule is that suggested for countries such as Britain where the Public Health medical services are well developed.

D

IMMUNIZATION SCHEDULE

Age	Visit	Vaccine	Time interval
2–6 months	1	Triple*	
	2	,,	4–6 weeks
	3	,,	4–6 weeks
7–12 months	4	Poliomyelitis (Oral)	
	5	,,	4–6 weeks
	6	,,	4–6 weeks
18–21 months	7	Triple	
2–4 years	8	Smallpox	
5 years	9	Dip./Tet. & oral polio.	
9 years	10	Dip./Tet. & smallpox	
15 years	11	BCG	

* Triple vaccine comprises diphtheria, pertussis and tetanus antigens.

Reference has already been made to the need for introducing and continuing other methods of disease prevention such as the provision of safe water supplies and the production of milk free from bovine tubercle bacilli; similarly the education of mothers regarding such matters as the preparation of safe milk feeds for artificially fed infants can do much to protect babies against infantile gastroenteritis which still makes a significant contribution to infant mortality rates in many developing countries.

Whilst such countries are faced with serious problems of health education Britain and other more sophisticated communities are similarly placed in regard to persuading mothers to have their children fully immunized; while it remains impossible to combine all vaccines then a minimum of eight visits are required for immunization purposes in the child's pre-school days and such a programme may become impracticable for the mother with a young family and a recent addition thereto.

Another reason for constantly reminding mothers of the need to have their children protected against certain infections is that the very success of active immunization procedures has led to the virtual elimination of some infections, e.g. diphtheria in this country so that the young mother of today is aware of the potentially lethal nature of such diseases only from family folk-lore.

CHAPTER 7

Serology

Serology is the study of the reaction between antigens and antibodies. Wide use is made of these reactions in diagnostic procedures.

(1) *Skin Tests*

There are several tests available where the principle is that a prepared antigen is injected intradermally in safe dosage and if the individual does not possess antibodies specific to the antigen, i.e. he has no specific acquired immunity, a skin response occurs, e.g. erythema with or without local oedema. Individuals who have acquired immunity will not show a response since the antigenic material is neutralized by the specific antibody in their tissues.

(2) *Laboratory tests*

In vitro methods have been devised which allow us to ob-
serve the specificity of reaction between antigens and their
antibodies. Obviously such tests can be used in two ways,
firstly, if we prepare antisera specific for particular bacteria
or bacterial components, these prepared or stock antisera
can then be used to identify bacteria isolated from pathological material. Alternatively, by maintaining stock strains of
fully identified bacteria, we can use these to detect specific
antibodies in samples of human serum.

Less commonly, laboratory animals can be used for the
in vivo demonstration of antigen-antibody reactions, e.g. in
determining whether a diphtheria bacillus isolated from an
individual is capable of producing exotoxin (see Chapter 13).

IN VITRO REACTIONS

AGGLUTINATION

Agglutination occurs where the test antigen comprises a suspension of intact bacterial cells and these clump together in
the presence of specific antibody and the aggregated cells
then form a deposit in the test-tube and the supernatant
fluid, which was originally turbid, becomes clear.

The phrase 'agglutination test' is almost synonymous with
the Widal test employed in detecting antibodies specific for
salmonellae because although agglutination techniques are
used to detect antibodies specific for bacteria in other
genera, the Widal test is far and away the most frequently
performed.

Widal test

As with most other *in vitro* serological tests the Widal test
is *quantitative*. A series of six tubes is set up so that each

contains an identical unit volume of patient's serum but in doubling dilutions from 1 in 15 to 1 in 480 and then to each tube is added a unit volume of stock bacillary suspension so that the final serial dilutions of patient's serum are from 1 in 30 to 1 in 960. A seventh tube containing only a unit volume of stock bacillary suspension and an equal amount of sterile physiological saline is included in the test to ensure that the stock suspension is not auto-agglutinable.

After thorough mixing, the contents of each tube are transferred by individual pipettes to agglutination tubes which are incubated at 37°C for 4 hours; thereafter the antibody titre of the patient's serum is taken as that in the tube containing the highest dilution of serum in which agglutination is noted.

If a formalized bacterial suspension is used in the test then the serum agglutinins which cause agglutination are flagellar antibodies, and similarly an alcoholized suspension detects somatic antibodies. Thus by parallel testing of separate aliquots of the patient's serum with formalized and alcoholized bacillary suspensions one can obtain additional information which aids interpretation of the result.

Interpretation of Widal test

Many individuals possess antibodies to certain salmonellae and yet have never suffered infection by such species; such *normal antibody levels* vary with the country in which the person lives and even vary in different parts of one country. In Britain the following titres are accepted as being within normal limits; for *Salmonella typhi* and *Salmonella paratyphi B*, H (flagellar) agglutination = 1 in 30 and O (somatic) agglutination = 1 in 50; for *Salmonella paratyphi A* and *C* both H and O reactions = 1 in 10.

Prior immunization with TAB vaccine can cause confusion in interpretation of results since titres, particularly of H agglutinins, may persist for many months after immunization. In such cases, doubt as to the significance of the result can often be resolved by repeating the test on a second serum sample taken 7–10 days after the first. If the titre of the second specimen is the same as that of the first then it is unlikely that the individual is suffering from active infection. However, even a definite rise in titre does not imply active infection since the rise may be caused by non-specific factors, e.g. a febrile, non-enteric condition, in which case the titre of the serum rapidly reverts to its original level when the non-enteric infection ceases.

Certain non-specific antigens, e.g. fimbrial antigens may to present in the test suspension and react with fimbrial antibodies in the patient's serum, often to very high titre, without significance. Obviously test suspensions must be devoid of such non-specific antigens.

PRECIPITATION

If instead of using intact bacterial cells we employ an extract of the cells as a colloidal solution and, in a tube, layer the antigen over its specific antiserum then precipitation occurs at the interface. The most common use of this method in diagnostic laboratories involves the determination of the group-specificity of β-haemolytic streptococci. After chemical extraction of the group-specific polysaccharide antigen from a culture it is layered into several tubes each containing equal volumes of various group-specific antisera and within a few minutes dense precipitation will occur at the interface of the tube containing the antiserum specific for the particular antigenic extract.

For research purposes gel-diffusion tests may be performed in Petri dishes or other suitable containers. Several variations of the basic technique are available but they all depend on using an agar gel which allows antigen and antibody to diffuse towards one another from wells or holes punched out of the agar and when specific components of each meet in the proper ratio, lines of precipitation appear in the agar. Such a method can be used to detect toxigenicity in diphtheria bacilli and can be used in place of *in-vivo* animal experiments (p. 89).

BACTERIOLYSIS

When, in the presence of normal serum complement, specific antibacterial antibody meets the bacterial cells which stimulated its formation the latter are lysed; this can be demonstrated *in vitro* by noting the elimination of turbidity in a tube test and also *in vivo* (Pfeiffer's reaction). In the latter instance a guinea-pig receives an intraperitoneal injection of cholera vibrios along with anticholera serum previously heated to destroy its natural complement. By withdrawing fluid from the peritoneum at 10 or 15 minute intervals one can note the vibrios undergoing lysis and within an hour or two none will be detectable. Here the natural guinea-pig complement has acted in concert with the injected antibody.

Complement is a group of proteins which occur naturally in the serum of individuals whether or not the person is immune to a disease or diseases. It deteriorates rapidly once serum has been withdrawn from the individual and can be speedily inactivated, without altering any antibody content, by heating the serum at 55°C for 30 minutes.

Bacteriolysis tests, as an indicator of specificity of antigen and antibody, are not used for diagnostic purposes but the analogous lysis of red blood cells (RBC) by specific haemolytic antiserum is made use of as an indicator system in complement-fixation tests; as in bacteriolysis, haemolysis of the target cell cannot be effected by complement alone or by the haemolytic antiserum if its natural complement has been inactivated. Hence, by mixing red cells with *heated* specific haemolytic antiserum one has a sensitized system which will remain unaltered in the absence of complement but which will show lysis of the RBC if complement is added.

COMPLEMENT FIXATION

If to a bacterial antigen one adds *heated* specific antiserum and a measured amount of complement from another source (usually guinea-pig serum) then when the antigen and antibody unite they bind or fix the complement, however the fixation of the complement is not visually detectable. One must then add a volume of sensitized RBC and since the complement is not available to complete the sensitized system there is no lysis of RBC and the complement-fixation test is positive, i.e. the heated serum contained specific antibody for the antigen in the original mixture. Conversely, if the stock antigen had no specific counterpart in the heated serum they would not unite and hence the complement would be free to complete the sensitized RBC system and when the latter is added lysis would occur, i.e. a negative complement-fixation test.

Although complement-fixation tests are widely used in other branches of microbiology, e.g. virology, they are not widely used in diagnostic bacteriology laboratories and indeed, the phrase 'complement-fixation test' is almost synonymous with the Wassermann test used in the serodiagnosis of syphilis.

The Wassermann test is performed quantitatively, thus one can follow the results of treating the disease since strongly positive reactions with patient's serum taken before therapy become progressively weaker as infection is eliminated. As with the interpretation of Widal test results various factors other than syphilitic infection may give a positive finding, e.g. in tropical countries the sera of patients suffering from yaws, which is a non-venereal treponemal infection, will react positively as will a proportion of cases of malaria.

In temperate zones false positive Wassermann reactions occur occasionally in patients suffering from collagen diseases and from certain infections, particularly of the respiratory tract. However, such reactions are short-lived and negative test results are found shortly after the causal, non-syphilitic condition is cured.

The sera of pregnant women occasionally show false positive Wassermann reactions and thus alternative tests are employed to confirm whether or not the individual is suffering from syphilis.

OPSONIZATION

Opsonins occur naturally in normal serum and they can alter the surface characteristics of bacteria which are then more susceptible to phagocytosis. Naturally occurring opsonins are thermolabile in contrast to opsonins found in the serum as a result of infection. These thermostable immune opsonins also facilitate phagocytosis of the bacterial cells which stimulated their formation.

Tests of opsonocytophagic activity in sera are technically difficult and not statistically reliable so that they are no longer in general use. They were performed by estimating the average number of bacteria ingested by phagocytes in the presence of a patient's serum and comparing this with the number of the same bacteria phagocytosed in a non-immune control serum.

NEUTRALIZATION TESTS

In cases where a suitable laboratory animal is susceptible to infection with a bacterium or alternatively reacts to a bacterial toxin, then the specific antibacterial antibody or antitoxin should protect the animal. In such tests a pair of animals is used and these should be as identical as possible, e.g. in age, sex, weight etc.; one is 'protected' (the control animal) by administration of antibody before it and its unprotected companion (the test animal) receive identical challenge doses of the antigen.

It may be that both animals show no response when challenged or both suffer ill effects. However, if the test animal shows pathognomonic evidence for the particular challenge material whereas the control animal suffers no effect, then we conclude that the antigen used to challenge the pair is specific for the antibody used to protect the control animal.

Obviously neutralization tests can also be used to detect 'unknown' antibody if one challenges with a known antigen and such tests can be made quantitative by using a series of biologically equivalent animals and, whilst maintaining a constant concentration of either antigen or antibody, one varies the concentration of the other reagent. This latter is the basis of assaying the potency of immunizing agents.

CHAPTER 8

Sources and Methods of Spread of Infection

An understanding of the sources and methods of spread of infection allows intelligent action to be taken, aimed at reducing the risk of infection in a susceptible individual and interrupting epidemic spread in a community.

The sources of infection for man are *other human beings, animals* and, in a few instances, the *soil*; human infections acquired from animal sources are called zoonoses.

Man as a source of infection

Most human infections are acquired from other humans and the source may be a sick person or a carrier. The severity of the infection in the sick individual does not necessarily indicate how dangerous he is to other people; indeed, a patient suffering from a mild infection often continues at work and can therefore disseminate the infecting organism to many more people than if his illness were severe enough to require isolation or bed-rest at home or in hospital.

Carriers may be defined as individuals who are found to be excreting pathogenic microorganisms but are not, at the time, suffering any disease; there are two kinds of carrier, the *convalescent* carrier is one who has recovered from an infection but continues to excrete the causal organism. Convalescent carriers are divided into temporary (transient) carriers and chronic (permanent) carriers depending on the duration of excretion after clinical recovery; the time interval after which a temporary carrier is regarded as becoming a permanent carrier varies with different diseases and the division is entirely arbitrary. In contrast to the convalescent carrier a person may be a *healthy* carrier, i.e. to the best of everyone's knowledge he has never suffered clinical infection caused by the organism he is excreting.

Human carriers are a frequent source of many infections and may be a greater risk to other men since their carrier state may not have been recognized. In certain circumstances and contrary to what might be thought, the chronic carrier is less of a risk as a source of infection than the temporary carrier, since the species which he is excreting may lose its pathogenicity over a period of time. For example, group-A streptococci produce reducing amounts of M antigen as carriage continues, and since this type-specific protein antigen is an important factor (perhaps the most important) in determining pathogenicity in such species we can explain why relatively few fresh cases can be traced to chronic carriers.

Animals as a source of infection

In most zoonoses it is unusual for the infected human being to act as a source for other men. Both domestic and wild animals act as sources and in either instance the animal may be sick or may be a carrier. Obviously there is an occupational risk of acquiring certain zoonoses, e.g. farmers and veterinary surgeons and abattoir workers have a continuing and closer involvement with many animals than the general population which may be protected by some legally required process, e.g. the setting up of cattle herds free from tuberculosis has brought about a drastic reduction in the incidence of human infection caused by bovine-type tubercle bacilli.

Soil as a source of infection

Certain species, particularly of the genus *Clostridium*, can be isolated from soil but since they are also excreted by man and animals we cannot be certain whether they are present in soil naturally or as a result of faecal contamination.

EXOGENOUS AND ENDOGENOUS INFECTION

When an individual suffers infection with organisms acquired from the sources mentioned above he is said to be infected from an *exogenous* source. Alternatively, *endogenous* infection may take place, i.e. the organism responsible for the disease has been living in or on the patient's tissues for some time. Reference has already been made (Chapter 4) to endogenous infection which can arise from essentially commensal species such as *Strept. viridans* or potentially pathogenic species which have been carried by the host for weeks or months before some alteration in the host-parasite relationship allows them to cause disease.

METHODS OF SPREAD OF INFECTION

More than 400 years ago Fracastorius's thesis *De contagione* detailed, in the first volume, his observations on the transmission of contagion from person to person; many of his ideas have been proved correct in the last few decades but we are still uncertain of the mechanisms involved in the spread of some infections.

Arthropod-borne blood infections

In such infections the causal microorganisms are present in the host's blood stream and are transmitted to a new host by blood-sucking arthropods such as mosquitoes, fleas, lice and ticks. In almost all such infections the microorganism can be spread only by its arthropod vector, but in certain cases the disease may affect tissues which allow transmission by other mechanisms, e.g. infection with bubonic plague is arthropod-borne to man from rats but occasional cases of plague also have a focus of infection in the lungs. Cases of pneumonic plague excrete the plague bacilli from their respiratory tract in large numbers so that the sputum of such cases is highly infectious to other men.

Venereal infections

The causative organisms of syphilis and gonorrhoea have very feeble powers of survival outside the human body, hence they must be transmitted speedily from host to host and in adult infection this is by sexual intercourse; there are of course, occasional instances of doctors and nurses being infected by careless examination of a syphilitic patient. Similarly, if the primary syphilitic lesion occurs on the lips or tongue of an individual who acquired the infection by abnormal sexual practices the causal organism can be transmitted to another person by kissing. Gonococcal infection of the eyes of babies can take place during birth if the mother is suffering from sexually acquired gonorrhoea. Similarly, occasional cases of gonococcal ophthalmia and vulvo-vaginitis can occur in young children infected from an adult attendant via sponges or damp towels.

Respiratory tract infections

In contrast with the spread of the above infections, a variety of methods of transmission occur in respiratory infections.

Organisms are expelled from cases of respiratory tract infection and by carriers of pathogenic species by spitting, nose-blowing and by their fingers; additionally they are disseminated by sneezing, laughing and coughing. Thus handkerchiefs, floors, furniture, clothing and other fabrics become contaminated with such secretions and these fomites can then act as vehicles of infection. Bacteria thus expelled can survive in the dust on surfaces for days or weeks and in the case of some pathogens even for months, e.g. tubercle bacilli, provided they are not exposed to direct sunlight.

Respiratory pathogens can thus be acquired in several ways, by *direct contact*, e.g. kissing, hand-shaking or by *indirect contact* where the new host transfers bacteria from clothing and other fomites with his hands into his nose or mouth.

Similarly *dust-borne* spread may occur when infected dust is made air-borne as when clothing is brushed or beds are made. Infected dust particles thus released may remain in the air for some time and be inhaled by the new host.

Finally, in addition to contact and dust-borne spread, a third method is possible namely, by *droplet spray*. Droplets of varying size are sprayed into the environment when we sneeze, cough, etc.; large droplets ($>$0.1 mm in diameter) immediately fall on to surfaces and contribute to infected dust. *Small droplets* ($<$0.1 mm in diameter) evaporate rapidly to form droplet-nuclei and because of their very small size ($<$10μ) remain air-borne and can be inhaled;

droplet-nuclei, however, rarely contain bacteria and only in some virus infections are they thought to play a part in transmitting pathogenic organisms.

Surface infections

Infection of skin, wounds and burns are acquired by mechanisms similar to those operating in the transmission of respiratory tract pathogens.

Alimentary tract infections

As with the spread of respiratory tract infections there are several methods of transmitting organisms causing bowel infection. Bowel pathogens are excreted in the faeces of cases and carriers and are, in general, less resistant to environmental agents than those causing respiratory tract infections although many survive for several weeks provided they are in moist surroundings.

Water-borne spread of infection occurs if an untreated water supply is fouled with excreta of cases or carriers of infection; this method of spread is classically associated with typhoid fever and cholera. The fouled water need not necessarily be imbibed since infection may be acquired from its use in preparing salads and other foods.

Hand-borne infection is very probably the principle method of spread in bacillary dysentery, particularly in well-developed communities with a safe water supply and methods of sewage disposal which prevent access of 'filthy, faecal-feeding flies' and other vectors to human excreta. A case or carrier will contaminate his hands while cleansing himself after defaecation and the pathogens can be transferred to toilet chains, wash-basin taps, door handles etc., thence to the fingers of another person and from there into his mouth. Similarly, hospital personnel may contaminate their fingers from bed-pans, soiled linen etc, and in addition to the risk of infecting themselves they may contaminate other fomites, e.g. drinking carafes, and thus transmit infection to patients.

Food-borne spread takes place when an infected individual is involved in food preparation or in other culinary activities where he may contaminate pots or pans in which the food is being processed. If foodstuffs are not protected from rodents and insects these, too, contribute to its contamination and subsequent infection by ingestion.

Finally laboratory acquired infection, either from pathological material or laboratory cultures, is a risk not only for the professional laboratory staff but to students in training. When screw-capped containers are opened aerosols are released and may be inhaled; fingers become contaminated from cultures and from working surfaces and transfer bacteria to the mouth; accidental self-injection with needle and syringe is not unknown, and obvious precautions must be constantly in force. It is essential that certain manoeuvres are carried out in specially constructed and ventilated protective cabinets; eating and smoking must be forbidden in the laboratory.

CHAPTER 9
Classification of Bacteria

Bacteria are primarily subdivided into *lower* and *higher* groups. *Lower bacteria* (Eubacteria) are unicellular and each cell is biologically independent; they never occur as sheathed filaments and are much more numerous than higher bacteria. Almost all bacteria which are pathogenic to man and animals belong to the lower group.

By contrast, *higher bacteria* (Actinomycetales) are fila-

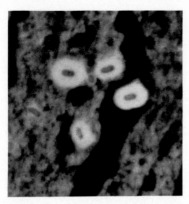

PLATE 1. Wet Indian ink film of a strain of *Klebsiella aerogenes* showing capsules and loose slime. × 1600.

Four bacilli can be seen and each is surrounded by a large capsule; the capsulate bacilli are on a background of loose slime which is partly infiltrated by carbon particles of the ink so that the slime is slightly darker than the bacterial capsules.

PLATE 2. Motility testing in semi-solid agar.

The tube of semi-solid agar on the left was inoculated by means of a straight wire to a depth of $\frac{1}{2}$ in. with a non-motile organism; the tube of semi-solid agar on the right was similarly inoculated but with a motile organism. After incubation at 37°C for 18 hr. it can be noted that the growth of the non-motile organism was restricted to the original inoculum track; in contrast, the motile organisms have spread throughout the medium and the inoculum track is not visible.

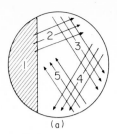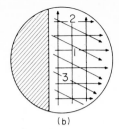

(a)   (b)

PLATE 3. Plating out.

The standard method of plating out is shown on the left; a reducing inoculum from the well area (1) through the series of strokes (2–5) is obtained by sterilizing the inoculating loop at each stage. On the right is shown the method of inoculating selective media; these media can be more heavily seeded and there is no need to sterilize the inoculating loop between the series of strokes.

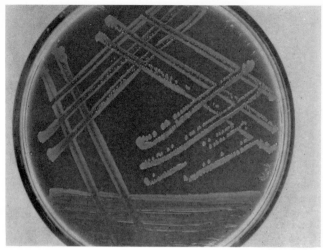

PLATE 4. Petri dish containing blood agar medium which was inoculated with a specimen of pus from a boil; after incubation overnight at 37°C there was a pure culture of staphylococci. The large size of the staphylococcal colonies contrasts with those of streptococci (Plate 7).

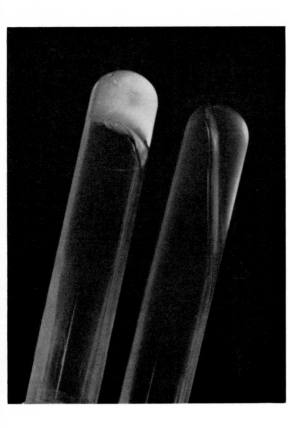

PLATE 5. Coagulase test.

The upper tube shows a positive test; the citrated plasma has been gelled by the coagulase produced by the pathogenic *Staphylococcus aureus* strain under test.

The lower tube shows a negative test; the contents are fluid.

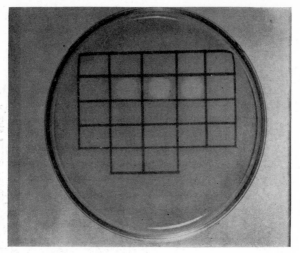

PLATE 6. Phage-typing of staphylococci.

A grid of 22 oblongs was marked on the outer surface of this Petri dish which contained digest agar and each oblong corresponds to the site of one phage preparation. The surface of the agar was flooded with a broth culture of coagulase-positive staphylococci and after the plate had dried 22 phage preparations were applied individually; after these preparations had dried the plate was incubated at 37°C for 18 hr. Examination of the plate revealed confluent lysis of the staphylococcal lawn at three sites corresponding to the areas of phage preparations 3B, 3C and 55.

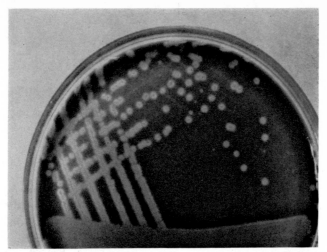

PLATE 7. This blood agar plate was inoculated from a throat swab and then incubated at 37°C overnight. The more obvious growth is that of *Strept. pyogenes* whose small colonies are surrounded by a zone of β-haemolysis (complete lysis of RBC) which is three times the diameter of the colonies; the less obvious colonies were of commensal *Strept. viridans*. The small size of both colonies contrasts with that of staphylococci (Plate 4).

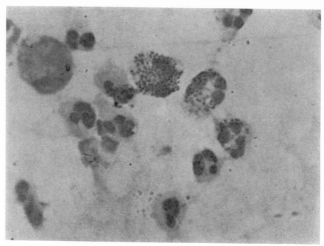

PLATE 8. Gram stained film of urethral discharge. × 1000.

Three of the polymorphonuclear leucocytes contain numerous diplococci; these were Gram-negative and their intracellular position is characteristic of pathogenic neisseriae. On cultivation a pure growth of oxidase-positive organisms was obtained and on biochemical investigation these proved to be *N. gonorrhoeae*.

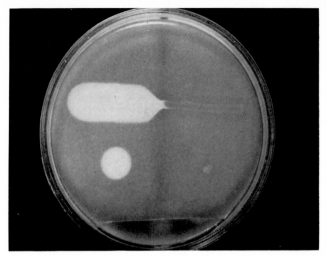

PLATE 9. Nagler's reaction. The right half of this egg yolk agar plate was smeared completely with three drops of *Cl. welchii* type-A antitoxin. After the plate had dried a culture of *Cl. welchii* was streaked from left to right and was also stabbed into the medium at two separate points, one on the left and the other on the right half of the plate.

The plate was then incubated anaerobically at 37°C for 20 hr. The organisms have grown on all of the inoculated areas and lecithinase activity has produced zones of opalescence on the antitoxin-free half of the plate but this activity was inhibited by the antitoxin on the right half.

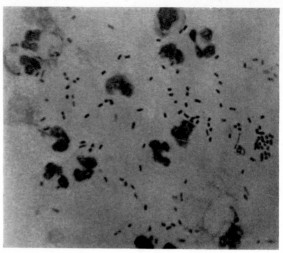

PLATE 10. Gram stained film of a specimen of urine from a case of cystitis. × 1000.

Numerous polymorphonuclear leucocytes, some of which are degenerate (pus cells) and many bacilli which were Gram-negative; the latter fermented lactose when grown on a Mac-Conkey plate and they had other characteristics typical of *Esch. coli.* Bacterial counts were of the order of 10^8 organisms per ml.

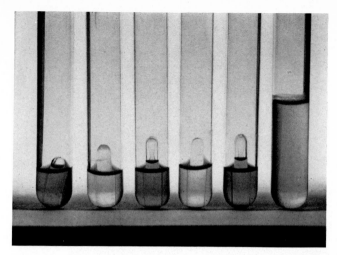

PLATE 11. The inoculum for this set of sugars and the tube of peptone water was a pale, lactose non-fermenting colony from a DCA plate. The sugars are respectively, from left to right, glucose, lactose, dulcite, sucrose and mannite. Glucose, dulcite and mannite have been fermented and with gas production as noted in the small Durham tubes. Indole was not produced in the peptone water culture but the organisms were motile; on this evidence the culture was potentially a member of the genus *Salmonella* and subsequent serological investigation proved the isolate to be *S. paratyphi B*.

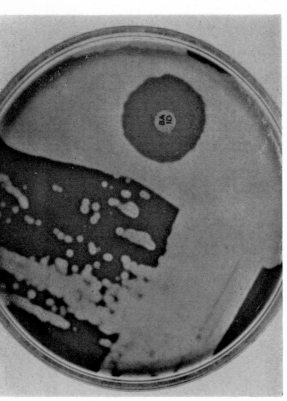

PLATE 12. This crystal violet blood agar plate was inoculated from a throat swab. A bacitracin disk was placed in the well inoculum area and after incubation for 18 hr. at 37°C a pure growth of β-haemolytic streptococci was noted; the growth was sensitive to bacitracin which indicates that the strain belonged to Lancefield's group A (*Strept. pyogenes*). Occasionally bacitracin-sensitive strains belong to group C or G.

mentous: some are sheathed and exhibit true branching with the formation of a mycelium; certain cells may have specialized functions, e.g. for reproduction, hence some interdependence occurs among higher bacteria. Only a very few are pathogenic to man; several useful antibiotics have been derived from higher bacteria, e.g. *Streptomyces griseus* produces streptomycin.

Lower bacteria are classified by their morphology and reactions to Gram's staining technique thus:

Cocci are globose cells, *Bacilli* appear as relatively straight cylindrical cells, *Vibrios* are definitely curved rod-shaped cells, *Spirilla* are spiralled non-flexuous rods and *Spirochaetes* are very thin, spirally twisted flexuous filaments (Fig. 4).

| Cocci | Bacilli | Vibrios | Spirilla | Spirochaetes |

FIG. 4. Primary classification of the lower bacteria.

Each of these morphological kinds of lower bacteria are further subdivided on the basis of staining reactions or more detailed investigations.

COCCI

Six main groups of cocci can be distinguished by noting their Gram staining reaction and the spatial relationship of cells, one to another, in each group. The groups correspond with biological genera and with the exception of one group, *Neisseriae*, all are Gram-positive (Fig. 5).

Staphylococci occur as irregular clusters of cocci, lacking

E

any orderly arrangement since successive cell divisions occur irregularly in different planes.

Streptococci adhere mainly in chains since consecutive planes of cleavage occur in the same axis.

Diplococci divide in a similar way to streptococci but adhere mainly in pairs and the cells in each pair are slightly elongate in the long axis of the pair. On *in-vitro* cultivation the cells become globose.

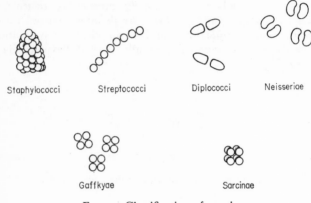

Staphylococci   Streptococci   Diplococci   Neisseriae

Gaffkyae   Sarcinae

Fig. 5.   Classification of cocci.

Neisseriae are Gram-negative; the cells adhere mainly in pairs and when seen in films of pathological material or from first cultures in the laboratory they are slightly elongate at right-angles to the axis of the pair.

Gaffkyae: division occurs consecutively in two planes at right-angles and hence they are seen as tetrads, i.e. flat plates of four cells.

Sarcinae are seen as cubes or packets of eight cells since division occurs consecutively in three planes at right-angles.

BACILLI

The primary subdivision of bacilli into groups is less exact than that for cocci and several of the groups described contain many biological genera which can be differentiated only by detailed study of their biochemical, serological and other properties (Fig. 6).

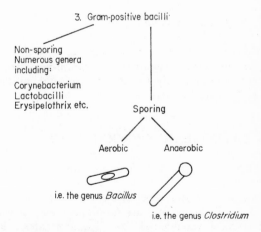

1. Acid-fast bacilli
   Mycobacteria

2. Gram-negative bacilli
   Numerous genera
   including:

   Salmonella
   Shigella
   Escherichia
   Pseudomonas
   Proteus
   Pasteurella etc.

3. Gram-positive bacilli

   Non-sporing
   Numerous genera
   including:

   Corynebacterium
   Lactobacilli
   Erysipelothrix etc.

   Sporing

   Aerobic   Anaerobic

   i.e. the genus *Bacillus*

   i.e. the genus *Clostridium*

FIG. 6.   Classification of bacilli.

Acid-fast bacilli, i.e. members of the genus *Mycobacter-ium*, are separable from other bacilli by virtue of their ability, once stained, to tolerate attempted decolourization with strong mineral acids.

Bacilli which are not acid-fast can be classified into those which are Gram-negative and others which are Gram-positive. Biological genera within the Gram-negative group include *Pseudomonas, Salmonella, Shigella, Escherichia, Proteus, Pasteurella, Brucella, Haemophilus* etc.

The Gram-positive group of bacilli may be subdivided into those genera which form spores, i.e. the genus *Bacillus* whose members are aerobic and the genus *Clostridium* whose members are anaerobic as well as numerous genera which cannot form spores, e.g. *Corynebacterium, Lacto-bacillus, Erysipelothrix* and *Listeria*.

Recognition of genera within the non-sporing Gram-positive bacilli and within Gram-negative bacilli demands more detailed study of morphological and other physiological attributes.

VIBRIOS AND SPIRILLA

Vibrios, definitely curved rod-shaped cells, and Spirilla, non-flexuous spiralled rods, are Gram-negative; species can be recognized only by detailed cultural, biochemical and serological methods.

PATHOGENIC SPIROCHAETES

These slim, spiralled, flexuous filaments are classified into three genera. Members of the genus *Borrelia* are larger than those of the other two genera and can be seen by the light microscope and are Gram-negative; they also have a larger coil wave-length $(2-3 \mu)$ and greater coil amplitude than the other genera (Fig. 7).

Treponemata are much finer with a shorter coil wave-length (1μ) of small amplitude; coils are more numerous than *Borrelia* and can be seen only by dark-ground microscopy or with the light microscope after staining by a silver impregnation technique.

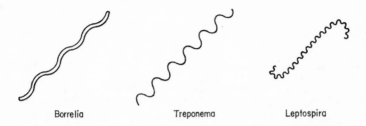

Borrelia   Treponema   Leptospira

FIG. 7. Classification of spirochaetes.

Leptospirae are even finer than *Treponemata* and the coils are so close—wave-length of 0.5 μ—that they can hardly be seen under dark-ground microscopy. One or both ends of the organism are often recurved on the body.

ACTINOMYCETALES

There are two genera of medical importance in the high bacteria.

Actinomyces are Gram-positive and mycelium-forming, usually also showing bacillary and coccal forms due to fragmentation and are anaerobic. The species infecting man grow in the tissues as colonies which become macroscopically visible.

Nocardia have many features in common with *Actinomyces* but they are aerobic and usually acid-fast.

CHAPTER 10
Gram-positive cocci

STAPHYLOCOCCI

Morphology

Each cell is approximately 1μ in diameter and spherical; they occur in irregular clusters of varying size and when grown in fluid medium clustering may be less obvious and chain-formation may be seen. Usually non-capsulate but small capsules are detectable on freshly isolated pathogenic strains. Non-motile, non-sporing.

Cultural requirements

Grow over wide temperature range, 10°–42°C. Optimum= 37°C. Aerobic and facultatively anaerobic and grow on the simplest of media.

Cultural appearances

Opaque, convex disks—2–4 mm after 24 hours incubation at 37°C, and pigmented. *Staph. aureus*=golden yellow, *Staph. albus*=white, *Staph. citreus*=lemon. Pigment formation is accentuated by growing on cream-agar (Plate 4, between pp. 56–7).

Biochemical activities

Earlier methods of differentiating pathogenic from commensal staphylococci, e.g. fermentation of carbohydrates and ability to liquefy gelatin, have been superseded by the much more reliable *coagulase test.*

Coagulase Test. Virtually all strains isolated from pathological material produce the enzyme coagulase whereas commensal strains rarely do so.

Five drops of an overnight broth culture of the strain to be tested are added to a tube containing 0.5 ml of a 1 in 10 dilution of citrated rabbit plasma, the diluent being sterile physiological saline. This tube and another inoculated with a known coagulase-producing strain are then incubated at 37°C and inspected hourly; clotting of the plasma occurs, usually within 4–6 hours if the strain produces coagulase. A negative control tube, i.e. one containing diluted citrated plasma to which is added 5 drops of sterile physiological saline, should be included with each batch of tests to ensure that the plasma does not undergo auto-coagulation (Plate 5, between pp. 56–7).

Serological characteristics

Strains can be typed using agglutination techniques with absorbed antisera but serological typing has been replaced by phage-typing which is more sensitive and more reliable.

Phage-typing. Phages are viruses which show a high degree of specificity for their host bacteria and produce lysis of the host cell. Thus if a plate of suitable medium is sown with a lawn of susceptible coagulase-positive staphylococci and suitable phage-preparations are individually applied, then after incubation there will be, at the site of implantation of certain phages, an absence of bacterial growth indicating that the particular phage was capable of lysing the staphylococci. By noting the pattern of growth-inhibition we can identify the phage-type of the staphylococcus (Plate 6, between pp. 56–7).

EPIDEMIOLOGY OF STAPHYLOCOCCAL INFECTIONS

Infections caused by staphylococci range from simple skin lesions, e.g. a furuncle, to more deep-seated conditions such as acute osteomyelitis. Sometimes there is an extension from the primary lesion which results in septicaemia or pyaemia with abscess formation in many organs and tissues.

The sources of *Staph. aureus* are other individuals who are either carriers or are suffering from infection; *endogenous* infection may occur in a person carrying the same phage-type in his anterior nares or perineum as that isolated from his lesion.

A carrier or case of infection will disseminate staphylococci into the environment from his clothing, skin squames and lesion and a susceptible host may become infected indirectly from fomites in the environment or directly by contact with the case or carrier. Such *exogenous* infection is the rule in hospital-acquired staphylococcal infection which usually involves wounds although staphylococcal pneumonia may also result.

Certain staphylococci produce an enterotoxin which, unlike other staphylococcal toxins, is heat-stable and can even survive boiling for short periods. If enterotoxin-producing staphylococci contaminate foodstuffs on which they can survive or grow, anyone ingesting such materials may suffer acute, toxic-type, food poisoning.

On occasion animals may be the source of infection.

Prevention

It will be obvious that exogenous infection in hospital could be controlled to a large extent if susceptible individuals were separated from carriers and cases of staphylococcal infection. Except in a few instances, such as patients being prepared for transplantation surgery, it is impossible in existing hospitals to ensure that sources of infection and susceptibles do not meet or share the same environment.

However, *contamination of the environment* can be reduced if intelligent measures are undertaken. For example, lesions must be covered with an impervious material and they should be dressed only in a dressings station and *not* in the open ward; wherever possible infected individuals should be nursed in isolation; carriage should be suppressed by applying suitable antimicrobial creams; and a continuing awareness of the part played by hands in spreading the organism from person to person should dictate a high level of personal hygiene. *Control of organisms in the environment* can be effected by oiling of floors, bedclothes, towels, pyjamas, etc. so that infected dust and other materials are trapped on the surface and are incapable of redissemination. Wet vacuuming of impervious floor surfaces is ideal but if the structure does not allow such treatment damp sweeping and dusting should be undertaken. Operating and dressings rooms should be properly ventilated with a positive pressure system so that air enters from outside the hospital and is not sucked in from corridors and wards.

Viability

Thermal death point (TDP)=62°C/$\frac{1}{2}$ hr. Survives outside the host more readily and longer than most other non-sporing bacteria; if protected from direct sunlight strains can exist in dust, bed-clothing, curtains, etc. for weeks or months.

Susceptible to disinfectants if these are used at correct concentrations.

STREPTOCOCCI

Morphology

Approximately 1μ in diameter, spherical and occurring usually in chains of varying length; when grown on solid medium some degree of clustering may be noted. Capsules occur in certain types when freshly isolated. Non-motile, non-sporing.

Cultural requirements

Similar to staphylococci but grow better on enriched medium, e.g. blood agar; narrower temperature range, i.e. 22°–42°C.

Cultural appearances

Circular, low convex disks, semi-transparent and 0.5–1 mm in diameter after 24 hr incubation at 37°C; their small diameter and the absence of pigmentation allow ready differentiation from staphylococcal colonies. Three types of response may be noted when streptococci are grown on blood agar but the colonies themselves are identical regardless of changes in the medium (Plate 7, between pp. 56–7).

α-Haemolytic streptococci are surrounded by a narrow halo of greenish discolouration of the blood agar medium; lysis of the red cells does not occur.

β-Haemolytic strains produce a much wider zone of complete haemolysis.

γ-Haemolytic (non-haemolytic) streptococci produce no alteration in the medium.

Since these three cultural types of streptococci differ in other biological characteristics and also in their clinical manifestations they require separate description.

α-HAEMOLYTIC STREPTOCOCCI (*STREPT. VIRIDANS*)

Biochemical activities

Although there have been many attempts to differentiate species within the group no practical classification has emerged. In identification, the most important point is differentiation from pneumococci which may appear very similar on cultivation on blood agar. This is most readily performed by testing for bile solubility or sensitivity to 'optochin' since *Strept. viridans* react negatively whereas pneumococci (q.v.) give positive reactions.

Serological characteristics

The fact that viridans streptococci are essentially commensal and do not spread epidemically may explain the relative lack of interest in their antigenic structure; thus what little is known, e.g. that they do not possess C-group antigens like β-haemolytic streptococci, is entirely academic.

Infections caused by *Strept. viridans*

These are always endogenous; such organisms are frequently found in carious teeth and are commonly present in periapical infections; otherwise they lead an essentially commensal existence in the buccal cavity of all persons. However, *Strept. viridans* is the most common cause of subacute bacterial endocarditis; this infection, which was invariably fatal before the introduction of antibiotics, occurs in individuals with predisposing cardiac lesions which may be *congenital*, e.g. patent ductus arteriosus, or *acquired*, e.g. rheumatic endocarditis. Such people are particularly at risk if they are in a poor state of dental hygiene when, even during normal mastication and more so during dental therapy, large showers of *Strept. viridans* enter the blood stream and may settle on the damaged heart valves or other areas.

Prevention

Individuals with heart lesions known to carry a risk of developing subacute bacterial endocarditis should be kept in an excellent state of dental hygiene and additionally must be given suitable antibiotic cover before, during and for 48 hrs after having dental treatment even of a minor nature.

β-HAEMOLYTIC STREPTOCOCCI

Biochemical activities

None of any practical use for purposes of identification.

Serological characteristics

Since these streptococci are the most common in infections of man and many animals, they have been subjected to rigorous antigenic analysis and the results have great importance in epidemiological practice. They are broadly subdivided into *serogroups* by testing for the presence of a particular group-specific carbohydrate antigen present in the cell wall. This carbohydrate or C-antigen can be extracted from the cell wall and its identity established by precipitation tests against group-specific antisera. Thus 15 serological groups A–Q (none designated I or J) can be recognized and group-A strains account for almost all human β-haemolytic streptococcal infections—such strains are named *Strept. pyogenes*.

Strept. pyogenes can be further classified into highly specific *serotypes* in precipitation tests with type-specific antisera prepared against their M-antigens. Each M-antigen is specific for its serotype strains, is protein in nature and is located at or near the cell surface. The M-antigens are also directly related to the pathogenicity of *Strept. pyogenes* since M-antibodies protect against infection with the homologous serotype; and strains which have lost the ability to produce M-antigen are non-pathogenic.

Thus *Strept. pyogenes* strains isolated from individuals involved in an epidemic situation can be serotyped and the source of the epidemic traced and removed.

Epidemiology of *Strept. pyogenes* infections

Strept. pyogenes is very commonly the cause of tonsillitis and pharyngitis; quinsy throat (tonsillar abscess) occurs as a complication. In individuals infected with strains which produce erythrogenic toxin and who do not possess immunity to the latter a characteristic punctate erythematous skin rash develops and they are then said to be suffering from scarlet fever. It must be emphasized that cases of scarlet fever are no more, and certainly no less, dangerous as a source of infection to others than cases of infection without the skin rash. *Strept. pyogenes* also causes other syndromes and frequently displays its ability to spread locally to adjacent tissues—e.g. adenitis, mastoiditis and otitis media may occur as complications of tonsillitis; it is also the cause of erysipelas and some cases of impetigo and occurs in wounds, burns and, until recently, was the commonest cause of puerperal sepsis.

Sources of infection are cases suffering from any one of the various syndromes produced by such organisms and carriers who most commonly are throat carriers; however, nasal carriage, although much less common, is more dangerous to susceptible individuals since it has been demonstrated that nasal carriers account for as many fresh infections in other individuals as do throat carriers. As with staphylococcal infections, cases and carriers of *Strept. pyogenes* extensively contaminate their clothing and the general environment and other people may become infected by direct contact, or indirectly via fomites. Airborne spread by dust particles or rarely by droplet spray also plays a part in spreading such organisms.

Acute Rheumatism

The causal role of *Strept. pyogenes* in acute rheumatic fever and other non-pyogenic sequelae of infection is now beyond doubt, although the mechanism whereby the organism produces the non-septic rheumatic complications some 2–3 weeks after the initial infection is still a source of debate and research. There is no association with any particular serotype and the risk of acute rheumatism, but certain subsidiary factors predispose, e.g. social circumstances, including poor nutrition, and geographic factors. It is usually stated that rheumatic fever and carditis are more common in temperate zones but this is not so and many cases occur in tropical countries. It would appear that altitude is an important determining factor since it is known that the primary streptococcal infection causing rheumatic fever becomes less common with increasing altitude.

Acute glomerulonephritis

This also occurs as a non-suppurative complication of *Strept. pyogenes* infection. In contrast with rheumatic sequelae glomerulonephritis is associated only with infections caused by very few serotypes with type-12 strains predominating; not all type-12 strains are nephrotoxic and other serotypes of which some strains may be nephrotoxic belong to types 4 and 25.

Prevention

Occasionally group-A streptococcal infection is endogenous in origin, particularly in erysipelas, but the majority of infections caused by *Strept. pyogenes* are acquired exogenously and prophylactic measures are the same as those for staphylococcal infection.

However, we have a most valuable means of preventing the spread of infection by *Strept. pyogenes*, i.e. the prompt and adequate treatment of cases with penicillin; this is possible since such organisms have remained eminently sensitive to this antibiotic and have not acquired the resistance which is so obvious in staphylococci, particularly in hospital strains of the latter.

Similarly, adequate treatment of streptococcal sore throat and other lesions with penicillin eliminates the risk of acute rheumatism and individuals who have a previous history of acute rheumatism should be protected against further streptococcal infection by continuous penicillin prophylaxis.

Unhappily penicillin treatment of streptococcal infection by nephrotoxic strains is not so efficient in protecting against acute glomerulonephritis. Even with early administration of penicillin in doses adequate to eliminate the causal organism one can expect little more than a 50% reduction in the incidence of acute nephritis.

γ-HAEMOLYTIC STREPTOCOCCI (*STREPT. FAECALIS, ENTEROCOCCI*)

Strains are usually oval in shape and occur in short chains. Unlike α- and β-haemolytic strains, *Enterococci* can grow on media containing bile salts, e.g. MacConkey's, and the colonies are minute, 0.5 μ in diameter, and usually magenta-coloured.

Biochemical activities

In addition to their ability to grow on bile salt media, *Enterococci* can also grow in the presence of 6.5% NaCl and ferment mannitol with gas-production, properties not shared by other streptococci. Four biochemical types, one of which occurs as three variants, may be recognized on the basis of gelatin liquefaction, fermentation of sorbitol and arabinose and other features.

Serological characteristics

All strains possess a C-antigen and belong to serogroup D.

Infections caused by Enterococci

Enterococci are essentially commensal in the intestine but can give rise to endogenous urinary tract infections, usually in association with Gram-negative bowel bacilli but occasionally as the sole cause.

Viability of Streptococci

It has already been noted that the temperature range over which streptococci will grow is narrower than that of staphylococci, their TDP is also lower, i.e. $54°C/\frac{1}{2}$ hr, except that *Strept. faecalis* again is exceptional and has a TDP of $60°C/\frac{1}{2}$ hr. The survival of streptococci outside the host is similar to that of staphylococci. *Strept pyogenes* is very much more resistant to crystal violet than are staphylococci and this fact allows us to make a blood agar plate selective for isolating *Strept. pyogenes* by incorporating a concentration of 1 in 500,000 crystal violet in the medium. Staphylococci on throat swabs will not tolerate such a concentration whereas *Strept. pyogenes* flourish.

PNEUMOCOCCI (*DIPLOCOCCUS PNEUMONIAE: STREPT. PNEUMIONAE*)

Morphology

Each cell is approximately 1μ in its long axis, slightly elongated (lanceolate) and cells adhere in pairs with their long axes in 'line-ahead formation'; short chains are also noted. Non-motile, non-sporing; capsules of varying size can be seen on strains freshly isolated from pathological material but capsulation diminishes and is eventually lost on continued *in vitro* cultivation. Similarly, laboratory cultures lose their lanceolate shape and become spherical after one or two subcultures.

Cultural requirements

Temperature range is even more restricted than streptococci—$25°$–$40°C$. Growth is enhanced if cultivation takes place in an atmosphere of 5% CO_2 and similarly the addition of 0.1% glucose to the medium is beneficial.

Cultural appearances

Colonies are similar in size to those of streptococci and are surrounded by a zone of α-haemolysis so that they may be confused with *Strept. viridans*, although pneumococcal colonies are plateau-shaped and not convex and ultimately the colonies develop an elevated edge and concentric ridges—the draughtsman colony.

Biochemical activities

Interest in these is restricted to tests which allow differentiation from *Strept. viridans*. Earlier tests have been superseded by determining the 'optochin' sensitivity of the isolate; pneumococci are extremely sensitive and viridans

F

streptococci are resistant to 'optochin'. The test is performed by sowing the strain under test on a blood agar plate and then placing a filter paper disk, impregnated with a 1 in 4000 aqueous solution of 'optochin' on the surface of the medium. Incubation for 18 hrs at 37°C will reveal a zone of inhibition of growth surrounding the disk if the isolate is a pneumococcus, whereas *Strept. viridans* will grow right up to the disk margin.

Another test which is equally reliable but technically tiresome is the bile solubility test; pneumococci are soluble in bile whereas *Strept. viridans* is not. One part of a sterile 10% solution of sodium taurocholate in normal saline is added to 10 parts of a broth culture of the organism under test and the mixture is then incubated at 37°C for 15 min. Lysis of the pneumococci is revealed by a clearing of the originally turbid mixture. The bile salt solution used must be crystal clear and the pH of the broth culture must be checked and if necessary adjusted to between 7 and 7.5, otherwise acid precipitation of the bile salt will result in turbidity.

Serological characteristics.

Serotyping of pneumococci is dependent on the highly specific capsular polysaccharides against which specific antisera can be produced for typing purposes. There are more than 75 serotypes of pneumococci. When a suspension of pneumococci is mixed with specific antiserum on a microscope slide and the preparation is viewed through an oil-immersion objective the capsules are sharply demarcated and appear swollen, whereas if mixed with heterologous antiserum the capsule is not visible. The use of the term 'capsule-swelling reaction' for the test is quite misleading since there is no increase in the size of the capsule which merely becomes obvious because of precipitation occurring between the capsular antigen and its specific antibody.

EPIDEMIOLOGY OF PNEUMOCOCCAL INFECTIONS

Lobar pneumonia is the disease which we primarily associate with pneumococci but they are also frequently involved as secondary pathogens in cases of bronchopneumonia where the primary infection is caused by a virus, e.g. in measles or influenza. Pneumococci are incriminated in a proportion of cases of acute pyogenic meningitis, either as a complication of pneumonia or as a primary illness. Individuals with traumatic or congenital defects in the skull may suffer recurrent attacks of pneumococcal meningitis unless the defect is repaired surgically. Pneumococci are also implicated in some cases of otitis media and conjunctivitis. Sources of pneumococcal infection are cases and carriers, and most pneumococcal infections are exogenously acquired. The modes of spread are the same as those of other organisms excreted from the respiratory tract. However, when pneumococci are involved as secondary invaders in bronchopneumonia, serotyping studies show that they are almost always the same type as those inhabiting the patient's upper respiratory tract and the super-infection is endogenous.

Prevention

General prophylactic procedures are the same as those used in the case of streptococcal infections. There are certain circumstances where a high incidence of lobar pneumonia, caused by only a few epidemic serotypes, can justify the use of active immunization with a polyvalent capsular antigen; in controlled trials a satisfactory degree of protection was noted. However, one could not justify immunization in normal communities when pneumococcal pneumonia is sporadic, responds readily to antimicrobial drugs and is not economically significant.

Viability

Although the TDP of pneumococci is lower than that of other Gram-positive cocci ($52°C/15$ min) and is more difficult to maintain in the laboratory, it has reasonable powers of survival outside the human host. This is evidenced by the fact that it can be recovered from the dust in the crevices between wood flooring boards some weeks after a room has been inhabited by a case or carrier.

CHAPTER 11
Neisseriae

Morphology

All members of the genus *Neisseria* appear as oval diplococci, each cell measuring approximately 1μ in its largest diameter. The longer diameters of each pair are parallel—'line-abreast formation'—and the opposed surfaces are flattened or concave. Non-motile, non-sporing. The two pathogenic members of the genus *N. meningitidis* (the meningococcus) and *N. gonorrhoea* (the gonococcus) become spherical on sub-cultivation; these pathogens are characteristically intracellular in films made from pathological material (Plate 8, between pp. 56–7); capsules are visible if specific antiserum is applied to wet preparations of the material as submitted to the laboratory or of colonies of the first laboratory isolates. The ability to form capsules disappears rapidly on *in-vitro* cultivation.

Cultural requirements

The pathogenic species are most fastidious, the gonococcus even more so than the meningococcus since the former will grow only on media containing blood or serum; temperature ranges for growth emphasize the more demanding nature of the gonococcus ($30°$–$39°C$) as compared with the meningococcus ($25°$–$42°C$). The growth of each species is enhanced by cultivation in an atmosphere of 5% CO_2.

By contrast the various commensal members of the genus grow readily on ordinary media and over a wide range of temperature—even at room temperature!

Cultural appearances

There is considerable variation in colonial morphology but the pathogenic species usually appear as small semi-transparent disks after 24hrs at $37°C$. Frequently on primary isolation the growth of gonococci may be slow and not evident for two or more days. Commensal species produce more opaque colonies which are frequently larger than those of the pathogenic neisseriae although some, particularly those of *Neisseria flava*, may resemble closely those of meningococci.

Biochemical activities

All members of the genus give a positive oxidase reaction, i.e. colonies show a rapidly deepening purple colour when a freshly prepared 1% solution of tetramethyl-p-phenylene-diamine hydrochloride is applied to them. N.B. The reagent is lethal to neisseriae if left in contact for more than 3–4 minutes, thus if colonies are to be subjected to further study they should be subcultured within the time limit stated.

Meningococci and gonococci can be readily differentiated from each other and from commensal neisseriae by performing fermentation tests using as substrates glucose, maltose and sucrose, but these must be contained in serum broth or agar and not peptone water so that the exacting pathogens will grow.

Meningococci ferment glucose and maltose only and gonococci only utilize glucose; commensal species either ferment all three substrates or alternatively do not attack any.

Serological characteristics

Meningococci belong to one of four groups, A–D and can be classified by agglutination tests with group-specific anti-sera. Gonococci have so far defied such exact antigenic characterization and seem to be serologically heterogeneous.

EPIDEMIOLOGY OF NEISSERIAL INFECTIONS

Commensal members of the genus live on mucous membranes including the buccal cavity, throat and urethra; they play no part in disease causation so far as we are aware.

Meningococcal infections

Meningococci are the most common cause of acute pyogenic meningitis which occurs sporadically and occasionally in epidemic form, particularly in military barracks or camps, or other establishments where a population is herded together in close proximity. Septicaemia, frequently running a chronic course and without meningeal involvement, is also associated with this organism and such cases may be labelled as pyrexias of uncertain origin until blood-culture is undertaken during an acute episode.

During epidemics of meningococcal meningitis serogroup-A strains predominate, whereas isolates from healthy carriers usually belong to groups B, C or D and it is thought that members of these groups are less pathogenic and/or less communicable.

Viability

TDP of pathogenic members is 55°C/5 mins or less and they are extremely susceptible to natural drying, sunlight and other environmental features either natural or artificially produced by man. Hence they are obligate human parasites. The speed with which meningococci die when discharged from the respiratory tract demands that their spread from human cases or carriers is fairly intimate and is probably by direct contact, e.g. kissing or perhaps by the recipient host breathing in large infected droplets of saliva when a case or carrier sneezes or coughs in close proximity. The exact method of spread is still in doubt.

Prevention

Apart from isolation of cases and elimination of over-crowding during epidemics coupled with good natural ventilation, no specific methods are available.

Gonococcal infection

Gonorrhoea is a venereal infection spread by sexual intercourse and the feeble viability of the organism outside the host belies the occasional claim that infection was acquired from toilet seats. Gonococcal ophthalmia neonatorum may result during the birth of a baby if the mother is suffering from sexually acquired infection and occasional cases of infection, either conjunctivitis or vulvo-vaginitis, still occur in children's institutions and are transmitted from an infected adult with poor personal hygiene via communal sponges and towels.

Prevention

Sexually acquired gonorrhoea is preventable by obvious methods; neonatal ophthalmia can be eliminated by ensuring that the mother is free from infection or alternatively by treating the potentially infected baby's eyes immediately after birth.

The other institutional types of non-sexually acquired infection can be prevented if nursing and medical attendants are free from infection or seek treatment and remain off duty until they are known to be cured.

THE GENUS VEILLONELLA

These are Gram-negative cocci, each only 0.3μ in diameter and appearing in clusters. They are anaerobic with an optimum temperature of $37°C$. Members of this genus are probably not pathogenic and are certainly commensal in the buccal cavity and intestines of man and animals.

CHAPTER 12
Acid-fast Bacilli

MYCOBACTERIUM TUBERCULOSIS

Morphology

Straight or slightly curved rods, variable in size but approximately $3\mu \times 0.3\mu$; non-motile, non-capsulate, non-sporing; Gram-positive, but difficult to stain by Gram's method; acid- and alcohol-fast. The five types of tubercle bacilli cannot be differentiated on morphological grounds.

Cultural requirements

Strict aerobes; the two types pathogenic for man, the human and bovine tubercle bacilli, have a fairly restricted temperature range for growth, i.e. 30°–41°C, and the optimum temperature is 37°C as is that of the murine type of bacillus. Avian tubercle bacilli have a slightly higher optimum temperature, 43°C, and the piscine or cold-blooded type grows best at 25°C. All types require a rich medium, e.g. Lowenstein-Jensen's (L–J), and even then visible growth does not appear for some *weeks*. Occasionally growth cannot be detected until eight or more weeks have passed, even with incubation under optimal conditions.

Further consideration is given only to the two types pathogenic for man but it should be noted that the three other types can be differentiated from each other and from human and bovine type bacilli by noting their cultural appearance and more particularly by demonstrating their differing virulence for various kinds of experimental animal.

Cultural appearances

On L–J glycerol-egg medium, human-type bacilli produce a dry, irregular, buff-coloured growth which is difficult to emulsify, whereas the growth of bovine-type bacilli is moist, smooth, white and readily suspended. The growth of the latter type on a glycerol-containing medium is less luxuriant than that of human-type bacilli.

Biochemical activities

What little is known of these has no significance to the clinical bacteriologist.

Serological characters

Four main sero-groups of tubercle bacilli are recognized, but the human and bovine types are antigenically indistinguishable so that sero-identification is of no value in the clinical bacteriology laboratory.

Animal pathogenicity

Guinea-pigs are highly susceptible to human and bovine type bacilli and are used frequently in diagnostic laboratory practice; in areas where cases of human disease are discovered at a very early stage guinea-pigs, inoculated with sputum etc, may show infection where attempted *in vitro* cultivation of the bacilli on L–J medium fails. On occasion there may be difficulty in deciding whether a particular isolate is of the human or bovine type and such doubts can readily be resolved by injecting the stain subcutaneously into a rabbit—'the bovine goes for the bunny'—i.e. bovine-type tubercle bacilli cause progressive disease within a few weeks and at post-mortem miliary spread of the lesions is obvious. By contrast the human-type bacilli do not cause infection in the rabbit, or at the very most a local lesion may result.

Epidemiology of Tuberculosis

Sources of infection are cases of pulmonary tuberculosis in other human beings and the new host is usually infected by breathing in bacilli lying in his environment. It is unlikely that droplet nuclei expelled by an infected individual contribute to the spread since the nuclei are too small to contain even one bacillus. Infection by inhalation usually results in pulmonary infection.

In communities where cattle suffer from infection with bovine-type bacilli and no attempt is made to eliminate these from the milk, man becomes infected by drinking the milk. Infection by ingestion most commonly causes intestinal tuberculosis.

However, both the human and bovine types have been incriminated regardless of the tissue or tract involved.

Primary tuberculosis in the lung usually gives rise to a small subpleural lesion with associated caseation of the hilar lymph glands. This pathological picture is known as the Ghon facus and almost invariably heals by fibrosis and calcification and the individual develops a positive tuberculin reaction. Post-primary tuberculosis may be endogenous since the bacilli which caused the Ghon facus can remain viable even in the healed, calcified lesion. As opposed to re-activation, post-primary tuberculosis may result from re-infection from another case. An important difference in the pathology of post-primary infection as compared with primary tuberculosis is that lymph gland involvement is less common in the former and the tissue lesions are progressive with cavitation; therefore the post-primary case is a greater danger to other people.

Prevention

The dramatic reduction in the incidence of human infection by bovine-type bacilli which has followed the control of bovine-type tuberculosis and pasteurization of milk underlines the importance of setting up and maintaining cattle herds which are free from infection.

Similarly, the use of BCG (Bacille Calmette-Guérin) vaccine to protect susceptible individuals has resulted in a significant reduction in the disease. In carefully controlled trials an 80% reduction in incidence was obtained and it was shown also that vaccination gave complete protection against the more serious types of infection, i.e. miliary tuberculosis and tuberculous meningitis. Another vaccine made from a murine strain, the vole vaccine, was also tested in these trials and was, at least, equally efficacious. Other prophylactic measures include the elimination of over-crowding and malnutrition which are known to increase the risk of infection; similarly regular radiographic examination, especially of those with an occupational risk, allows earlier diagnosis and hence more rapid cure. Tuberculosis is one of the few diseases where adequate disinfection, e.g. by formalin vapour, of rooms and furnishings, used by patients must be carried out before occupation by other people.

OTHER MYCOBACTERIA

(1) Intermediate group

Several types which, in so far as pathogenicity for the human host is concerned, are intermediate between tubercle bacilli and the saprophytic acid-fast mycobacteria have attracted attention within recent years. These are associated with chronic ulceration of the skin and in the case of one organism, *Mycobacterium balnei*, there is evidence that infection was acquired in swimming pools and entered through skin abrasions. *Mycobacterium ulcerans*, first described twenty years ago in Australia, produces essentially similar lesions. *M. balnei* and *M. ulcerans* can be differentiated from each other and from tubercle bacilli by various methods and in particular have optimum temperatures for growth between 30°–33°C and grow poorly, if at all, at 37° —the optimum for tubercle bacilli.

(2) Anonymous group

Members of this group have been isolated alone and in pure cultures from cases of 'pulmonary tuberculosis' but are readily differentiated from human and bovine type bacilli. In many instances members of the anonymous group are undoubtedly responsible for the tissue lesions whereas in others they have acted as secondary invaders in a true tuberculous infection. Unlike tubercle bacilli all anonymous mycobacteria can form the enzyme arylsulphatase and they are subdivided into *photochromogens* (which produce pigment only when exposed to the light), *scotochromogens* (which can produce pigment even when growing in the dark) and non-chromogens (which occur as colourless colonies or have very slow pigment production on exposure to light). Various other biological properties allow detailed differentiation of strains.

(3) Saprophytic and commensal group

Numerous species can be recognized; they grow very rapidly and on ordinary media. The only members of the group with significance to the medical bacteriologist are firstly *Mycobacterium smegmatis* which occurs commensally in the smegma of men and women and may therefore be present in specimens of urine submitted from suspect cases of genito-urinary tuberculosis. Its presence in such specimens need not cause confusion, since although it is acid-fast it is usually readily decolourized by alcohol. It is also less tolerant of the chemical methods used in concentrating specimens for cultivation and animal inoculation; even if it escapes the concentration technique its rapid growth on other media and lack of virulence for guinea-pigs allows ready differentiation from tubercle bacilli. Similarly saprophytic mycobacteria often inhabit water pipes and taps so that one should ensure that water used for preparing reagents for Z–N staining and for washing films during such staining is free from such species.

MYCOBACTERIUM LEPRAE

This organism is regarded as being the causative agent in human leprosy although the association is suspected solely on the fulfilment of the first of Koch's postulates. *M. leprae*

has not been isolated and attempts at experimental reproduction of the disease by implanting lepromatous material
from a case into animals have not resulted in established
disease.

One therefore must rely, for diagnostic purposes, on the
microscopic demonstration of acid-fast bacilli in material
taken from a suspect lesion; the leprosy bacillus is not as
acid-fast as the tubercle bacillus and the strength of sulphuric acid used in attempted decolourization should be 5%
and not 20% as is used for demonstrating acid-fastness in
tubercle bacilli.

The organism would appear to have a low infectivity and
may spread by direct contact with a very long incubation
period but we are still ignorant of the exact mechanism of
spread; opinions vary as to the prophylactic value of BCG
vaccination.

CHAPTER 13
Aerobic Gram-positive Bacilli

CORYNEBACTERIA

Several species are commensal in man and others are pathogenic for certain domestic animals but only one member of
the genus, *Corynebacterium diphtheriae*, is pathogenic to
man and gives rise to diphtheria.

C. DIPHTHERIAE

Morphology

Size and shape are variable, particularly in films made from colonies after laboratory isolation. Expansion of one pole to give a club-shaped organism is frequently seen. Incomplete separation of cell walls during division leads to 'Chinese-letter' arrangement. More easily decolourized in Gram's staining method than most other Gram-positive organisms. Volutin granules, demonstrable, e.g. by Albert's staining technique, appear in preparations made from colonies on rich media such as serum-agar and are scanty or absent in films of colonies from selective tellurite-containing media. Non-capsulate, non-motile, non-sporing.

Cultural requirements

Aerobic, wide temperature range for growth—20–40°C—optimum 37°C. Grows on ordinary media but best on media containing serum or blood.

Cultural appearances

On serum media, e.g. Loeffler's, growth is very rapid and after 12–18 hrs at 37°C circular grey colonies can be noted; these have a regular edge but the colonies increase in size after 24–48 hrs incubation and the edges are then crenated. Volutin granules are abundant in films made from such media; since many other organisms will also grow in Loeffler's serum medium it is customary to inoculate a selective medium containing tellurite at the same time as the serum medium.

Tellurite media suppress the growth of most other organisms likely to be present in the throat-swab specimen and also are slightly inhibitory to *C. diphtheriae*.

Incubation, therefore, should be continued for 48–72 hrs if no growth is apparent at 24 hrs.

Three colonial types of *C. diphtheriae* can be distinguished on tellurite media and these are termed *gravis*, *mitis*, and *intermedius* variants since respectively they are generally associated with severe, mild and moderate clinical illness; all types reduce the tellurite in the medium so that the colonies are grey or black. No detailed description of the colonies is offered since these vary and in any case recognition is a matter of practice and expertise. In general, *gravis* and *mitis* types are similar in size but the former often has an irregular crenated edge with radial striations, whereas *mitis* type colonies are usually convex and circular in outline. *Intermedius* types produce colonies which are smaller in size and frequently have a flattened border which is circular; colonies of this type are black as opposed to the slate-grey colour of the other types. The few organisms other than diphtheria bacilli which grow on tellurite media can, by the expert, be differentiated from them on colonial appearance.

Biochemical activities

All the colonial types ferment glucose but only *gravis* strain ferment glycogen; they can be further differentiated by noting their varying haemolytic activities against ox and rabbit red cells and in tube haemolysis tests *gravis* strains lyse only rabbit cells, *intermedius* strains have no effect on either type of red cell whereas *mitis* strains lyse both ox and rabbit cells.

Serological characteristics

The three bio-types of *C. diphtheriae* are readily separable by antigenic analysis and each can also be subdivided by tests with agglutinating antisera. Diphtheria bacilli produce a powerful exotoxin which spreads throughout the patient's body and causes general toxaemia with obvious clinical effects on the circulatory and nervous systems and on renal tissues. The few strains of each bio-type which are non-toxigenic cannot cause diphtheria.

Animal pathogenicity

Several laboratory animals are susceptible to diphtheria exotoxin but usually guinea-pigs are used to determine whether an isolate is toxigenic. Two guinea-pigs are used in toxigenicity testing and one animal, protected by intra-peritoneal administration of diphtheria antitoxin, acts as a control. Both control and test guinea-pigs have their abdomens shaved before 0.2 ml of a 12 hrs culture of the isolate is injected intradermally. Toxigenic, i.e. virulent bacilli will produce a local erythematous area which within 32–48 hrs will become necrotic, but in the control animal no reaction will occur. If neither animal shows a reaction the strain is non-toxigenic and if both animals react then the organism is not a diphtheria bacillus. Toxigenicity can also be determined *in vitro* by a gel-diffusion technique.

Epidemiology of Diphtheria

Sources of infection are cases of the disease and carriers of virulent bacilli; both nose and throat are sites of carriage. Direct contact, e.g. kissing of the infected child by parents and siblings before he is removed to hospital, was undoubtedly a likely method of spread in the days when diphtheria was rife in this country and when parents were well aware that the chances of the sick child surviving were not high. Spread by fomites, e.g. drinking mugs or school pencils, has been incriminated and inhalation of dust particles contaminated with diphtheria bacilli was probably the most important mechanism of transmission.

G

Prevention

The remarkable reduction in incidence following on mass immunization of the community with toxoid preparations early in the Second World War underlines the high degree of protection afforded by active immunization. There is a fall-off in the proportion of people receiving such protection when the medical profession does not publicize the importance of protection to the individual and thus the community.

Isolation of cases, immediate emergency protection of contacts who are not known to be protected and eradication of the carrier state all play a part in attempting to prevent the spread of diphtheria in an epidemic or potentially epidemic situation.

DIPHTHEROID BACILLI

As already stated, several members of the genus *Corynebacterium* are commensal in man.

Corynebacterium hofmannii

This occurs in the throat and like other commensal species does not produce exotoxin and is not pathogenic. More uniform in its morphology than *C. diphtheriae*, it rarely possesses volutin granules and with simple staining, e.g. with methylene blue, an unstained central bar can be seen and this is characteristic. Colonial appearances differ from those of diphtheria bacilli and Hoffman's bacillus does not ferment glucose.

Corynebacterium xerosis

This is commensal in the conjuctival sac—morphologically similar to *C. diphtheriae* but can be differentiated from the latter by its ability to ferment sucrose and in lacking patho-genicity for the guinea-pig.

Several other diphtheroid bacilli can be identified but these two are the most commonly encountered.

THE GENUS BACILLUS

Only one member of this genus is pathogenic to man and to certain animals, i.e. *Bacillus anthracis*, the causative organism of anthrax.

BACILLUS ANTHRACIS

Morphology

Large $(4\text{--}8\mu \times 1\text{--}1.5\mu)$, rectangular, Gram-positive bacilli with a tendency to form long strands or chains. Capsulate in the tissues and body fluids of the infected host, non-motile and forms spores when existing outside the host; characteristically the spore is oval in shape, central in position and does not project beyond the confines of the vegetative cell.

McFadyean's reaction

This reaction, which is characteristic of anthrax bacilli when capsulate, can be noted by microscopic examination of films of peripheral blood taken from an infected animal. The film is fixed by 1 : 1000 mercuric chloride for 5 mins and then polychrome methylene blue stain is applied for 15 secs; in such a preparation the disintegrated capsules appear as amorphous, irregular masses of heliotrope debris among which are lying large blue bacilli.

Cultural requirements

Organism grows readily on all ordinary media and over a wide temperature range (12°–45°C) with an optimum for growth of 35°C. Although the vegetative cells will tolerate wide variations in their gaseous environment spore formation occurs only under aerobic conditions.

Cultural appearances

Colonies have a 'medusa-head' appearance, i.e. a wavy margin resembling locks of hair and each colony represents a continuous thread of bacilli; white, opaque and like ground-glass, some 3–4 mm in diameter after 24 hrs incubation.

Biochemical activities

Anthrax bacilli ferment a variety of sugars but such tests are not employed in identifying the organism because other more striking characteristics allow easy recognition.

Serological characteristics

Several antigenic materials are now recognized and of these the exotoxin is the factor which causes death of the host. Exotoxin was discovered only a few years ago and until then it was thought that death was caused 'mechanically' by the massive septicaemic proliferation of the bacilli which block capillary blood vessels. Two somatic antigens, one a protein and the other a polysaccharide, can also be demonstrated as well as a capsular, polypeptide antigen. The latter can stimulate the formation of capsular antibody which however is without any protective effect against infection.

Animal pathogenicity

Guinea-pigs are extremely susceptible and subcutaneous injection of freshly isolated capsulate strains or of pathological material, e.g. blood from a cow dying of anthrax, results in death in 24–48 hrs; the pathological picture is identical with that occurring in a naturally infected animal. There is a marked inflammatory response at the injection site and the local lesion is teeming with bacilli which can also be found in large numbers in the heart-blood and all organs; the spleen is particularly involved, being grossly enlarged and friable—hence the name 'splenic fever'.

Epidemiology of anthrax

Anthrax is primarily an infection of domesticated herbivorous animals but all mammals are susceptible in varying degrees; man only becomes infected by contact with sick animals or products of animals which have died from the disease.

The anthrax spore produced by vegetative cells outside the host tissues is the infecting agent in both animal and human infection. In animals, infection results usually from ingestion of the spores from contaminated pasturages where they can survive for many years. In man, anthrax is commonly a localized infection of the skin and subcutaneous tissues ('Cutaneous anthrax') and the spore gains access through a surface abrasion; certain groups of people have an occupational risk, e.g. farmers and veterinary surgeons.

Infection may result from inhalation of the spore when it contaminates wool and pulmonary anthrax in man frequently becomes septicaemic; gastro-intestinal anthrax in man (clinically identical to that in animals) occurs only in the most undeveloped societies where the carcase of an anthrax-infected animal may be used as food—in such circumstances epidemics are the rule and the mortality rate is very high.

Prevention

The carcases of animals dying from anthrax must be carefully disposed of either by cremation or deep earth burial in a quick-lime pit—specific procedures are detailed in the Anthrax Orders and have the aim of limiting spore formation and the dissemination of spores. Post-mortem examination of experimentally infected laboratory animals must be made with great care to protect the operator and the total environment of the animal house. One form of cutaneous anthrax, 'Hide-porters disease', was associated with lesions across the shoulders or neck of dock porters who humped bales of contaminated hides whilst unloading ships; this was brought under control by restricting the importation of hides to one British seaport, Liverpool, and providing facilities for the mechanical handling of potentially infected material until this had undergone a rigorous disinfecting process. More recently, cutaneous anthrax has occurred sporadically and also in epidemic form in association with the use of bone-meal fertilizer containing anthrax spores and elimination of this source is under active consideration.

Pulmonary anthrax (Wool-sorter's disease) in Britain has been controlled in wool factories by carefully planned local exhaust ventilation which removes wool fibres from the air and prevents their inhalation by workers.

Gastro-intestinal anthrax in humans can be readily prevented by ensuring that man cannot have access to anthrax carcases. Various kinds of vaccine are available for active immunization of animals; spore vaccines prepared from non-capsulate virulent strains have been used successfully in many countries but are not used to protect humans since they are considered not to be sufficiently safe. Individuals following occupations with a high risk of infection can be actively immunized with a vaccine made from the somatic protein antigen; its protective value against cutaneous anthrax is satisfactory.

OTHER MEMBERS OF THE GENUS BACILLUS

There are numerous saprophytic species in this genus and these are collectively referred to as anthracoid bacilli. They are ubiquitous and are frequently encountered as contaminants in the laboratory. Their occurrence need not create any difficulties in differentiation from anthrax bacilli since, unlike the latter, they do not give a McFadyean reaction and are non-pathogenic for guinea-pigs when administered in doses similar to those of *B. anthracis*. Many of the saprophytic species are motile and give marked haemolysis on blood agar in comparison with the feeble haemolytic activity of the anthrax bacillus.

Lactobacilli

Members of this genus are essentially commensal but one, i.e. *Lacto. acidophilus* plays a part in promoting dental caries; lactobacilli are an important part of the normal flora of the human intestinal tract and also abound in the adult vagina and here they have a protective value since they produce lactic acid from the glycogen secretions, thus creating an acid environment which prevents the growth of many other, potentially pathogenic organisms. Lactobacilli are remarkably pleomorphic and can occur as long slender rods or even as cocco-bacilli with much intermediate variation in size and shape; they are non-motile and non-sporing. They are aciduric and selective media, e.g. tomato peptone agar, have a low initial pH of 6 or less; most species are micro-aerophilic or anaerobic on primary isolation and colonial appearances can be as variable as microscopic morphology.

Dental caries

Lactobacillus acidophilus undoubtedly participates in the

production of dental caries, which is of multifactorial origin. The part which they play has been a source of confusion for many years but experimental proof is quite definite in showing that animals fed on a cariogenic diet do not develop caries until lactobacilli are added to the diet, when the disease develops rapidly.

CHAPTER 14
Anaerobic Gram-positive Bacilli

Anaerobic spore-bearing Gram-positive bacilli are collectively members of the genus *Clostridium*. There are numerous species and the majority lead a saprophytic existence and play an important rôle in the decomposition of dead animal and plant life; some occur commensally in the intestinal tract of man and animals and a few produce potentially lethal diseases in man—botulism, tetanus and gas gangrene.

CL. BOTULINUM

Morphology

Straight-sided rods with rounded ends, measuring approximately $4\mu \times 1\mu$. Non-capsulate, motile with a peritrichous flagellar distribution, spores are oval, subterminal and projecting.

Cultural requirements

Strict anaerobe, temperature range for growth is 20°–37°C, optimum is 35°C; grows on ordinary media.

Cultural appearances

Considerable variation but colonies often have an irregular edge and those of toxin-producing strains are semi-transparent whereas the colonies of non-toxigenic sporing variants are opaque.

Biochemical activities

The six antigenically distinct types show variation in their saccharolytic and proteolytic activities, but details of these are not necessary for our present purposes.

Serological characteristics

The six types (A–F) produce antigenically distinct exotoxins and each of these neurotoxins can be neutralized only by its own antitoxin and this allows us to establish the type of any strain isolated.

Animal pathogenicity

Guinea-pigs and other laboratory animals when challenged by inoculation or by feeding with cultures are susceptible and clinically suffer an illness similar to that in man; animals may be protected with specific antitoxins so that we can demonstrate the type of neurotoxin produced by an isolate.

Epidemiology of Botulism

Although every type has been incriminated, in human cases types A, B and E are most frequently involved. The organisms are saprophytic and can be isolated from soil, fruit, vegetables, etc.; man is affected by ingesting foodstuffs in which the organisms have had time to form their exotoxin and almost all reported outbreaks have been associated with *preserved* foods of various kinds. Botulism is essentially an *intoxication* and not an infection—secondary cases are rare, if they occur at all.

Prevention

The food *industry* is well aware of the danger of botulism and precautions are taken to ensure that foodstuffs are heated sufficiently to destroy even the spores of *Cl. botulinum*; *home* preservation of foods is much more risky as is reflected by the seasonal incidence in North America where most cases occur during the winter months when such home-preserved foods are consumed more frequently and in larger quantities than at any other season.

Antitoxin may be administered to individuals who may have eaten food which is suspect; active immunization is possible with a mixed toxoid preparation but is not justified in countries where the disease is rare.

CL. TETANI

Morphology

Straight, rod-shaped and measuring approximately $5\mu \times 0.5\mu$; motile with peritrichous flagella, non-capsulate. Spores are spherical, terminal and projecting and for this reason the organism is referred to as the 'drum-stick bacillus' but it is not unique in this regard.

Cultural requirements

Strict anaerobe and even established, *in vitro* cultures die rapidly on exposure to the normal atmosphere; wide temperature range for growth 14°–42°C, optimum 37°C. Grows on ordinary media but growth is enhanced if blood is incorporated.

Cultural appearances

Discrete colonies are not usually seen and growth is characterized by a fine diaphanous film from the edges of which extend numerous long branching projections. Non-motile variants give rise to isolated colonies which rarely exceed 1 mm in diameter and are transparent with an entire circular edge devoid of any projection.

Biochemical activities

No saccharolytic activity; slight proteolytic action as shown by minimal digestion of meat particles in cooked-meat broth.

Serological characteristics

Ten types of *Cl. tetani* can be distinguished on the basis of flagellar antigens but all ten serotypes produce an identical exotoxin; certain strains in each serotype may be non-toxigenic.

Animal pathogenicity

In addition to man, several animals are naturally susceptible to tetanus and in the laboratory mice are employed for diagnostic purposes. As in similar animal tests with other bacteria or their toxins, a pair of animals is used in each test and one of the pair is protected with tetanus antitoxin before both are challenged. Unprotected mice inoculated intramuscularly in a hind leg show evidence of tetanus within a few hours; the tail stiffens and the inoculated limb becomes paralysed. Paralysis then becomes generalized and tetanic spasms occur on the slightest stimulus.

Epidemiology of Tetanus

In contrast to *Cl. botulinum* which can be isolated from virgin soil and therefore leads a truly saprophytic existence, it is not certain that tetanus bacilli are also saprophytic; the soil population may be derived from faecal contamination by animals and man and certainly the organisms are most prevalent in soil which has been manured.

Almost invariably tetanus spores enter a wound by means of soil contamination; occasionally catgut, made from sheep's intestines, is inadequately processed and has acted as a source of infection in surgical wounds. The mere presence of spores in a wound does not invariably result in clinical tetanus since their germination is dependent on various factors; devitalized tissue, the presence of foreign material such as soil or clothing and the coexistence of aerobic organisms are factors which contribute to a reduction in oxygen tension around spores and allow them to germinate as well as encouraging the growth of the vegetative cells which emerge. Toxin-producing strains present in a wound remain restricted to the site of wounding and the clinical picture is caused by the effects of the neurotoxin which diffuses to the nervous system.

Tetanus does not spread directly from man to man and when epidemics occur they are due to several individuals being infected from a common source.

Prevention

Emergency prophylaxis in the non-immunized individual depends on early and thorough surgical treatment of the wound combined with the administration of tetanus antitoxin. Alternatively, penicillin can be given prophylactically if the patient has a history of hypersensitivity to horse serum.

Long-term protection, particularly to groups at special risk, e.g. farm workers, depends on active immunization with tetanus toxoid; the basic course of immunization comprises three injections each of 0.5 ml of toxoid with a 6-weeks interval between the first two, the third injection being given some 6–12 months after the second dose.

The reduction in incidence of tetanus neonatorum in more primitive societies is essentially dependent on education in safe methods of dressing the umbilical stump, since there is little doubt that a dressing of cow-dung is the usual vehicle for the tetanus spores.

Obviously, cat-gut, dressings, umbilical-cord powder and other materials used in wound dressing and surgical procedures must be free from tetanus spores.

CL. WELCHII

This species is by far the most commonly incriminated in cases of gas-gangrene in man, either as sole agent or in combination with other members of the genus, e.g. *Cl. oedematiens* and *Cl. septicum*.

Morphology

Bacilli have rounded or square ends and measure approximately $5\mu \times 1\mu$. Unlike all other clostridia, *Cl. welchii* is *non-motile*. Forms capsules in animal tissues but not on *in vitro* cultivation. Spores are subterminal, oval and non-projecting.

Cultural requirements

Anaerobic but not as strictly so as *Cl. tetani*; grows rapidly at $37°C$ and particularly if a fermentable carbohydrate, e.g. glucose, is incorporated in the culture medium.

Cultural appearances

While there is some variation in colonial morphology, colonies are usually large (3–5 mm), semi-opaque with an entire edge, and on blood agar β-haemolysis is usually evident.

Biochemical activities

Ferments a variety of sugars, e.g. glucose, lactose and maltose, with production of gas. In litmus milk medium *Cl. welchii* produces acid with clotting of the medium and gas, produced from fermentation of the lactose, disrupts the clot resulting in the 'stormy clot reaction'.

Serological characteristics

Five types (A–E) of *Cl. welchii* can be identified according to the types of major lethal toxins they produce; type A strains are most commonly pathogenic for man and these produce only the alpha toxin, whereas types B–E, which are usually associated with disease in various animals, produce in addition one or more of the beta, epilson and iota toxins.

Nagler's Reaction. Alpha toxin is an enzyme, lecithinase C, and its antitoxin is highly specific. When a strain producing lecithinase is grown on an egg yolk medium a zone of opalescence surrounds the growth and this precipitation can be inhibited if alpha antitoxin is present. This is known as Nagler's reaction and is performed by spreading the antitoxin over one half of the surface of the medium before streak inoculation of the organism is made, so that it will grow on both halves of the medium. It is important to inoculate the plate from the untreated half to the antitoxin half, otherwise antitoxin might be carried over to the normal surface (Plate 9, between pp. 56–7).

Animal pathogenicity

Strains vary greatly in their pathogenicity; pathogenic strains injected intramuscularly kill guinea-pigs within 24–48 hrs. An hour or two after injection oedema occurs in the injected leg and gas formation allows elicitation of crepitation in the tissues; oedema and tissue destruction spread rapidly and the organisms can be recovered from the heart-blood 8–12 hrs after injection. Animals injected with specific antitoxin prior to challenge with a culture show no ill effects.

Epidemiology of Gas-gangrene

Cl. welchii is a normal inhabitant of the large intestine of animals and man, thus manured soil or other materials in which spores may exist, e.g. clothing, act as a plentiful reservoir. Thus, as in the case of tetanus, a wound may be contaminated and gas-gangrene might result. Detailed study of wounds during the Second World War revealed that many wounds in which pathogenic species were present healed without any evidence of gas-gangrene and in others, although the disease was present in the injured muscle, it did not invade healthy muscle in the same limb and this localized infection is referred to as anaerobic cellulitis. Finally, classical gas-gangrene may occur when organisms display their full invasive ability and spread very rapidly into healthy muscle which then undergoes toxic necrosis.

Of the several other members of the genus which are incriminated in gas-gangrene, *Cl. oedematiens, Cl. septicum* and *Cl. bifermentans* occur most commonly; these strains can be differentiated from each other and from *Cl. welchii* biochemically and by serological methods.

Prevention

Prophylaxis is primarily dependent on *thorough surgical treatment* of a wound soon after injury. Ancillary preventive measures include the administration of suitable antibiotics and although the infection is rare in civilian life, the administration of antitoxic serum should not be withheld in severe wounds which might be liable to gas-gangrene. Polyvalent serum is given if the organisms have not been isolated and monovalent serum when specific identification has been made.

CHAPTER 15
Gram-negative Bacilli, I

The family *Enterobacteriaceae* contains many genera and apart from the fact that some are motile and others are not *they are morphologically* indistinguishable.

Morphology of enterobacteria

Size variable, but approximately $3-5\mu \times 0.5\mu$, relatively straight cells with rounded ends, some species may be capsulate and some possess fimbriae. None produces spores.

ESCHERICHIA COLI

The majority of strains are motile.

Cultural requirements

Grows over a wide temperature range—15–41°C, optimum 37°C; also tolerant of atmospheric conditions. Grows well on ordinary media.

Cultural appearances

On nutrient agar, colonies are large, 2–4 mm in diameter after 18 hrs at 37°C; opacity of colonies varies with different strains but all are convex and with an entire edge. On MacConkey's medium colonies are similar to those on agar but are rose-pink since lactose is fermented; *Esch. coli* do not grow well on deoxycholate-citrate-agar (DCA) and colonies are small and pink.

Biochemical activities

Like most other enterobacteria, *Esch. coli* has a wide range of activity on many substrates. However, the bacteriologist engaged in diagnostic work makes use of only a few tests which allow him to differentiate *Esch. coli* strains from microscopically similar species which, in the same site, may be pathogenic, e.g. *Esch. coli* almost invariably ferment lactose whereas salmonella and shigella strains cannot do so; similarly indole production by *Esch. coli* differentiates them from salmonella species.

Serological characteristics

Strains can be serotyped for epidemiological purposes by identification of somatic and flagellar antigens with relevant specific antisera. However, most interest centres around the K (Kapsule) antigens, which are present in certain strains which can cause gastroenteritis. These K

H

antigens of which three types, L, A and B, can be differentiated, reside on the surface of strains either in a recogniz-
able capsule or as an envelope which is too fine to be seen.

Epidemiology of Esch. coli infection

Esch. coli is essentially a commensal of the intestinal tract
of animals and man and thus it is widely distributed in the
environment.

Endogenous infection of the urinary tract is the usual
type of infection caused by such strains but they are also involved in peritonitis, appendix abscess and wound infections
either alone, or with other organisms, e.g. enterococci
(Plate 10, between pp. 56–7).

Exogenous infection of the urinary tract also occurs and
the organisms are usually introduced during catheterization
either on inadequately sterilized instruments or because of a
breakdown in aseptic technique.

Gastroenteritis resulting from infection with enteropathogenic strains of *Esch. coli* is not restricted to any age
group but is most commonly seen in epidemic form in
infants; the source of such strains may be a carrier or a case
of infection and the higher incidence among artificially-fed
babies emphasizes that spread from the source is often by
means of contaminated milk-feeds.

Prevention

In exogenously acquired infection of the urinary tract,
wounds, etc. prophylaxis depends on adequate sterilization
of instruments, dressings, and other materials and thorough
aseptic procedures in operating and dressings rooms and in
the ward.

Terminal heat-treatment of feeds for infants can eliminate the major source of gastroenteritis but, in addition, the
mother and hospital personnel must be rigorous in their

personal hygiene. In the domestic situation the mother should be advised to prepare a feed immediately before use and should be instructed in the aseptic handling of the feeding bottle and the raw materials in preparing the feed.

PROTEUS

The morphology and cultural requirements of members of this genus are very similar to those of *Esch. coli*. Almost all strains are motile.

Cultural appearances

On nutrient agar, isolated colonies are only occasionally seen and these are similar to those of *Esch. coli*; usually *Proteus* species swarm over the surface of the medium in successive waves. Since non-motile variants do not swarm the phenomenon is obviously associated with motility but its exact cause is not known.

Swarming can be inhibited if the strain is grown on MacConkey's medium or if one of several agents, e.g. 1 in 500 chloral hydrate, is added to the medium. Swarming has a nuisance value to the bacteriologist since the sheet of growth may cover colonies of other significant organisms.

Biochemical activities

Four biochemical types of *Proteus* (see Table overleaf) can be recognized and all are readily differentiated from other enterobacteria by their ability to produce urease; although some other members of the family *Enterobacteriaceae* can decompose urea, none can do so as quickly as the *Proteus* species.

BIOCHEMICAL TYPES OF PROTEUS

	Pr.vulgaris	*Pr.mirabilis*	*Pr.morganii*	*Pr.rettgeri*
Mannitol fermentation	−	−	−	⊥ or +
Maltose fermentation	−	+	−	−
Indole production	+	+	−	+
Citrate utilization	−	−	−	+

Serological characteristics

All four biochemical types can be further subdivided by agglutination reactions into many groups and serotypes depending respectively on their somatic and flagellar antigens but serotyping is not performed routinely.

The specific nature of the reaction between an antigen and its antibody has been referred to in Chapter 5 but one of the few exceptions in specificity of such reactions concerns certain types of *Proteus* which have antigens identical with some rickettsiae. These types are used as antigens in agglutination tests with serum from patients suspected of suffering from rickettsial infection, e.g. the Weil-Felix reaction in cases of typhus fever.

Epidemiology and prevention of Proteus infections

With the exception that Proteus species do not cause gastro-enteritis, the sources and methods of spread of infection caused by them are identical to those of *Esch. coli.*

SHIGELLAE

All members of the genus *Shigella* are non-motile but otherwise are identical in morphology and cultural requirements with *Esch. coli*.

Cultural appearances

These resemble those of *Esch. coli* except that they do not ferment lactose and the colonies are therefore pale or colourless; members of one group, i.e. *Sh. sonnei*, are late lactose fermenters thus, after incubation on MacConkey's or DCA medium for 24 hrs or more, their colonies have a light pink appearance.

Biochemical activities

Four biochemical groups of shigella can be recognized:

TABLE 2

	Glucose	Lactose	Mannitol	Indole
Sh dysenteriae	⊥	—	—	∨
Sh. flexneri	⊥	—	⊥	∨
Sh. boydii	⊥	—	⊥	∨
Sh. sonnei	⊥	(⊥)	⊥	—

With the exception of certain strains of *Sh. flexneri* serotype 6, no gas is produced in fermentation reactions.

Serological characteristics

Each biochemical group is antigenically distinctive and with the exception of *Sh. sonnei* each can be subdivided into several serotypes and in the case of certain *Sh. flexneri* strains subtypes can also be recognized.

Epidemiology of Bacillary dysentery

Members of the genus *Shigella* are parasites of man and of a few higher apes. Infection is by ingestion, and to some extent the severity of clinical illness is associated with the particular group to which the organism belongs, tending to be more serious when *Sh. dysenteriae* strains are involved whilst in the case of *Sh. sonnei* infection, a mild, short illness is the rule.

In countries or areas with inadequate sewage disposal systems, flies and other insects may feed on human excreta and then soil foodstuffs, but in more sophisticated countries insects do not play a rôle and methods of spread from a human case or carrier is 'hand-to-mouth'. Direct contact as in hand-shaking and indirect contact via toilet fixtures and door handles is the usual method of transmission.

In Britain, where for many years Sonne dysentery has predominated and increased in incidence, the disease can be used as an index of personal hygiene; the mild nature of Sonne dysentery allows many cases to remain ambulant and thus to act as peripatetic disseminators.

Prevention

In Britain it is probable that improved personal hygiene, particularly hand-washing after defaecation and before handling food, could greatly reduce the spread of Sonne dysentery. Even then, the use of communal towels would allow the transmission of the organisms from person to person so individual towels made of paper or other materials should replace roller-towels. Of course, in areas where flies have access to faeces physical and chemical fly-control methods should be employed to protect foodstuffs.

Perhaps one of the factors contributing to the increasing incidence of Sonne dysentery in the past was the fact that, until recently, no method of typing the serologically homogeneous *Sh. sonnei* was available. Colicine typing allows the recognition of at least 17 epidemiologically significant types and this typing method should encourage more detailed investigation of outbreaks. No vaccines are available for active immunization.

SALMONELLAE

With two exceptions all of the 1,000 or more serotypes of salmonellae are motile; their morphology and cultural requirements are very similar to those of *Esch. coli.*

Cultural appearances

Salmonellae resembles closely the pale colourless colonies of shigellae.

Biochemical activities

Salmonellae are variously active on many substrates and with the exception of *S. typhi* fermentation is accompanied by gas production. For preliminary identification it is usual to show that glucose, dulcitol and mannitol are fermented and that a strain does not utilize lactose or sucrose and that indole is not produced (Plate 11, between pp. 56–7).

Subsequent confirmation of identity is by serological methods but some serotypes, e.g. *S. enteritidis,* can be subdivided into epidemiologically significant varieties by extended tests of their biochemical ability.

Serological characteristics

A salmonella isolated from pathological material can be allocated to one of several *groups* by identifying its *somatic* antigens with group-specific antisera; the *type* identification within any one group depends on the use of highly typespecific *flagellar* antisera. There are potential complicating factors in both grouping and typing procedures. Firstly, a few salmonellae, e.g. *S. typhi*, have a third antigen, namely the Vi (virulence) antigen which occurs as a surface coating and masks agglutination with somatic antisera; before the somatic antigen can be recognized a suspension of the Vi-possessing strain must be boiled for 1 hr and then washed by centrifuging.

Secondly, the flagellar antigens of any one species may be present in either or both of two phases; phase 1 antigen is specific for its own type species, whereas phase 2 antigen is non-specific and is shared with many species; thus an isolate must be in the specific phase 1 state before it can be typed.

Isolates which are found to be in the non-specific phase, i.e. the *majority* of the culture population possesses phase 2 antigen, can usually be obtained in the specific phase by using Craigie's tube method, which employs semi-solid agar to which has been added phase 2 antiserum (1 in 50). Within the main container is placed an open-ended piece of glass tubing which projects above the surface of the medium; when phase 2 organisms are inoculated in the agar *within* the glass tubing and incubated they are agglutinated by the phase 2 antiserum and immobilized. Only phase 1 organisms, which are in the minority in the inoculum, remain motile and can thus spread from the inner tube and escape from its lower end into the medium outside the tube. Thus by taking sub-cultures from the surface of the medium *outside* the inner tube one can harvest specific phase bacilli.

Epidemiology of Salmonella infections

Salmonellae can cause either the enteric fevers, i.e. typhoid and paratyphoid fever, or food poisoning and these differ in many respects.

Enteric fevers

These are caused by *S. typhi* and the three types *A*, *B* and *C* of *S. paratyphi*. Human cases and carriers are the only sources of infection and in the case of typhoid fever the spread from a human source is classically by means of water supplies; however, foodstuffs also act as vehicles of infection in the enteric fevers. *S. paratyphi B* is the type most commonly encountered in Britain.

After ingestion the organisms reach the small intestine and pass by lymphatic spread to the mesenteric glands where they multiply and then, via the thoracic duct, invade the blood stream. Only at this stage does the patient become a source of infection and increasingly so as the bacteraemic phase comes to an end. During the bacteraemic phase many tissues are invaded including the liver, gall bladder and kidneys; from the gall bladder a second invasion of the intestine takes place and on this occasion obvious pathological effects occur with particular involvement of the lymphoid tissue, the bacilli being excreted in large numbers in the faeces. Similarly, if the organisms become localized in the kidney they will be excreted in the urine.

In approximately 3% of patients excretion continues after clinical recovery and although carriage is often temporary it may continue indefinitely. In chronic carriage the bacilli are usually present in the gall bladder or less commonly in the urinary tract. Cases of enteric fever which are recognized and treated in isolation are not as great a danger to the community as are carriers. Apart from the risk of contaminating water supplies, carriers may contaminate foodstuffs with bacilli which are present on their hands and many foodstuffs have thus been incriminated as vehicles of infection.

As in bacillary dysentery, insects may act as vectors of spread if they have access to infected faeces.

Salmonella food poisoning

In comparison with the enteric fevers the incubation period
of salmonella food poisoning is short (1–2 days as against
10–14 days), the clinical illness is brief and lasts only a week
or less, infection is often localized to the gut, and bacteraemia is not a constant finding and man is not the only
source of infection. Domestic animals, e.g. cattle, sheep
and pigs, and also fowls can suffer infection and also act as
carriers; similarly rodents are a constant potential source of
infection.

A carrier rate of 1% has been found in healthy beef
cattle and sheep and it has been shown that this rate increases as animals are moved from farms to abbatoirs,
especially if the transit period is lengthy and/or the animals
are overcrowded in trucks and are not given a plentiful
supply of food and water.

Carcases can obviously become contaminated at any
stage during marketing or during preparation for human
consumption in butchers' premises, hotel and other kitchens
either from other carcases, from human cases or carriers or
from the excreta of rats and mice.

Intensive farming techniques undoubtedly increase the
risk of animals transmitting infection to each other and
hence increase the danger to the human population. This
situation is even more unhappy if the animals are fed with
foodstuffs containing antibiotics in an endeavour to increase their beef yield and hence their market value. Thus
salmonella strains in such animals can and do develop
resistance to antibiotics which otherwise might be useful in
treating the person ultimately suffering infection.

Milk and milk products may also be a source of infection,
as can hen and duck eggs; if these are pooled or used in
making custards etc., which will be communally consumed then several individuals will be at risk. On the other
hand, many cases of individual infection are probably un-
detected when an egg is eaten by one person.

Prevention of enteric fevers

Measures which are generally enforceable include the provision and maintenance of safe water supplies and adequate water-borne sewage systems, supervision of workers in water-works and in the food industry, protection of foodstuffs from insects and bacteriological control of imported foodstuffs. Similarly the prompt isolation and treatment of all cases is required and known carriers must not be employed in any situation where they might contaminate water or food.

Specifically, people at special risk, e.g. troops or travellers in areas where the disease is endemic, can be offered a reasonable degree of protection by active immunization with phenolized TAB vaccine; for people living in Britain, only those living with chronic carriers and laboratory personnel working with specimens or live suspensions need be offered immunization.

Prevention of Salmonella food poisoning

Human cases and carriers must obviously be prevented from working in any situation where they might contaminate articles of food and drink, including such articles intended for animal consumption; infected animals must speedily be withdrawn from the herd or flock and killed.

Abattoirs, wholesale and retail butchers' shops and other premises where food is prepared or cooked should be maintained in a clean state and such premises should be rodent-proof and fly-proof.

Wherever possible food should be eaten immediately after cooking but when this is impracticable it must be stored in refrigerators so that any organisms which have survived can at least be prevented from multiplying. The education of food-handlers in the need for a high level of personal hygiene and in methods of handling foodstuffs so that contamination is minimal is a very important aspect of prophylaxis. No protective vaccines are available.

KLEBSIELLAE

Morphology

Size, shape and response to Gram's staining method are identical to other enterobacteria; non-motile, non-sporing but usually are capsulate *in vivo* and *in vitro* and, in addition, members of this genus produce large quantities of loose slime.

Cultural requirements

Similar to those of genera described above.

Cultural appearances

Non-capsulate, non-slime-forming mutants produce colonies similar to those of *Esch. coli* but normally the colonies are very viscid due to the extracellular materials which they produce.

Biochemical activities

These are not normally tested in diagnostic laboratories where identification is usually made on colonial appearances; urease is produced but *much more slowly* than by *Proteus* species and the lack of motility of *Klebsiellae* prevents confusion with Proteus organisms.

Serological characteristics

Somatic antigens in klebsiellae are usually masked by the capsular (K) antigens and/or the slime (M) antigens; in any one strain the M and K antigens are identical and by employing capsular antisera one can recognize at least 70 serotypes. The method of typing is the same as that used for pneumococci—the so-called capsule-swelling reaction.

Epidemiology

Although certain serotypes of *Klebsiella* are causally related to infections of the urinary tract and wounds and one particular species, *Kl. pneumoniae*, carries a high mortality rate on the infrequent occasions when it causes pneumonia, it must be appreciated that the majority are saprophytic or commensal.

Whilst *Klebsiella* strains are incriminated in urinary tract infections, they feature particularly in cases where there are predisposing gross lesions and are much less common where such lesions do not exist; infection may be endogenous or exogenous.

Apart from the occasional severe case of pneumonia, members of the genus are also found in some cases of paranasal sinusitis and acute otitis media, but it is not yet clear whether such organisms have been derived from an exogenous source or have existed as commensals in the upper respiratory tract prior to showing their pathogenic potential.

In addition to the genera mentioned above, there are several others in the family *Enterobacteriaceae*, e.g. *Cloaca, Hafnia,* but these are not considered here since they are rarely encountered as human pathogens, or perhaps one should say they are not recognized!

PSEUDOMONAS PYOCYANEA (*Ps. aeruginosa*)

Morphology

Size variable but approximately $1.5\mu \times 0.5\mu$, motile with polar flagella or often a single flagellum; non-capsulate, non-sporing.

Cultural requirements

Aerobic—most strains strictly so; wide temperature range for growth 5°–43°C, optimum 37°C. Grows on ordinary media.

Cultural appearances

Colonies measure 2–4 mm and are low-convex, circular and often have an irregular spreading edge; cultures have a characteristic musty odour. Many strains produce pigments, pyocyanin and fluorescin, and their colonies have a greenish-blue fluorescent appearance. Some strains produce pyorubrin which endows their colonies with a reddish-brown colour. A significant minority of strains are unable to produce pigments even when special media are used for their detection. On MacConkey's medium colonies are similar in appearance but are pale since lactose is not fermented.

Biochemical activities

With the exception of glucose none of the sugars usually employed in diagnostic laboratories is utilized. The oxidase test (Chapter 9) is positive within 30 secs of the reagent being applied to a culture of *Ps. pyocyanea,* and although a few other Gram-negative bacilli also react with the test solution they do so much more slowly and rarely within 2 mins. Thus the test is valuable in identifying strains which do not produce pigments.

Serological characteristics

Several attempts have been made to serotype strains on the basis of somatic and flagellar antigens but none of the schemes has been widely accepted. The epidemiological tracing of strains has been simplified by the introduction of type-identification by pyocine production—a method similar to that used in typing strains of *Sh. sonnei*.

Epidemiology of Ps. pyocyanea infections

Although there are more than 140 species in the genus *Pseudomonas* only *Ps. pyocyanea* is pathogenic to man; some other members are pathogenic for plants and animals but the majority lead a saprophytic existence.

Ps. pyocyanea also occurs widely in nature and is present as a commensal in the intestine of man and can be isolated from healthy skin. In fulfilling a pathogenic rôle it is often associated with pyogenic cocci or with one or other of the enterobacteria, e.g. *Esch. coli*, and its importance as a human pathogen has increased, particularly in the hospital environment since the introduction of antibiotics, to almost all of which it is naturally resistant.

Whilst infection may be endogenous, exogenous sources are other human cases and to a lesser extent carriers; methods of spread are probably mainly by indirect contact and via dust. The very wide temperature range for growth explains why *Pseudomonas* strains living in drains, moppails, sinks, etc., can multiply very easily so that the entire environment becomes contaminated. Similarly many pieces of apparatus in hospital, e.g. respirators, once contaminated, can act as vehicles of infection.

The ability of the organism to resist many disinfectants accounts for some hospital epidemics since a particular contaminated lotion or solution may be applied to many patients from a communal container.

Infection may be localized to a wound, burn, or to the urinary tract but in patients already debilitated, e.g. with multiple injuries, fatal septicaemias occur frequently.

Prevention

Prevention of exogenously acquired infection demands that patients who are suffering infection should be nursed in isolation and that patients who are particularly susceptible to infection should also be nursed in strict isolation. Even so, the individual is at risk unless instruments, respirators, etc., are carefully and completely sterilized before use. With strict adherence to aseptic procedures the risk of infection will be reduced but even in isolation the ubiquitous nature of *Ps. pyocyanea* dictates the possibility of infection from fomites. Bacteriological monitoring of wards, etc., has revealed that even when all obvious vehicles have been eliminated bacilli can still be detected and may be present in the most mundane situations and transmitted to patients, e.g. strains will flourish in the earth of potted plants, and if the latter are watered by being left overnight in a partly filled bath, *Ps. pyocyanea* can be recovered from the bath after it has been emptied and cleaned!

CHAPTER 16
Gram-negative Bacilli, II

HAEMOPHILI

H. INFLUENZAE

Morphology

Usually appear as small cocco-bacilli, only $1.5\mu \times 0.5\mu$, non-motile and non-sporing; capsulate in young cultures. In old

cultures, and also in CSF from cases of meningitis caused by
H. influenzae, the bacilli often appear as very long slender
filaments.

Cultural requirements

Aerobic, temperature range for growth is 23°–39°C, optimum 37°C. Will not grow on nutrient agar and requires
both haematin (X factor) and nicotinamide-adenine dinu-
cleotide (V factor) to be present before it will grow; these
are supplied by incorporating blood in the medium.

Cultural appearances

On blood agar colonies are small (1 mm) and transparent
with an entire edge. If the essential X and V factors are
liberated from the blood cells in the agar plate by heating,
thus producing what is known as a chocolate agar plate,
growth is enhanced. Similarly, if grown in mixed culture
with another species which produces V factor in abund-
ance, e.g. staphylococci, *H. influenzae* colonies, in the
vicinity of colonies of these other species, are much larger
in size. This is known as *satellitism*.

Biochemical activities

Strains vary in their fermentative capacities and in any
case their activities are feeble.

Serological characteristics

Freshly isolated capsulate strains can be allocated to one of
six serotypes (a–f) by noting, in capsule-swelling tests,
which of six type-specific antisera reacts with the strain.

I

Epidemiology

H. influenzae is a strict parasite and man is the only host; both capsulate and non-capsulate strains are found in the throats of healthy people. Although the method of spread from person to person is not known, it is probably by direct droplet spray since the organism is feebly viable outside the host's tissues. As well as fulfilling a commensal role, *H. influenzae* can act as a secondary invader in the respiratory tract following infection with other microorganisms. Influenza bacilli are commonly found in chronic bronchitis and since clinical improvement follows antimicrobial therapy which eliminates them they are assumed to participate in the disease process.

H. influenzae is incriminated as a common cause of pyogenic meningitis, especially in children of pre-school age, and in this and other infections serotype b strains are isolated more frequently than strains of the five other types.

Contrary to earlier suggestions, influenza bacilli are not causally related to influenza, although they may act as secondary invaders in this and other virus infections of the respiratory tract.

Prevention

Only in people at special risk are any prophylactic steps taken. In those suffering from chronic bronchitis long-term treatment with a tetracycline, particularly during the winter months, reduces the incidence of acute bronchitic episodes.

OTHER HAEMOPHILI

H. aegyptius (Koch-Weeks bacillus) causes an acute, highly infectious type of conjunctivitis; in its morphology, cultural requirements, etc., it is very similar to *H. influenzae*. It can, however, be differentiated from the latter by serological tests.

H. ducreyi is the cause of chancroid, a venereal infection, and although similar in size to the influenza bacillus it frequently appears in pairs and short chains. It is extremely difficult to grow although it does not require the V factor.

BORDETELLAE

BORD. PERTUSSIS

Morphology

Similar to *H. influenzae*.

Cultural requirements

Very similar to *H. influenzae* but on subculture can be grown on ordinary media; for primary isolation, however, must be grown on Bordet-Gengou medium. Does not require X or V factors.

Cultural appearances

Growth may not be apparent until after 2–3 days incubation at 37°C. Colonies are minute, 1–2 mm in diameter with a circular edge and are dome-shaped and glistening and have been likened to a bisected pearl.

Biochemical activities

Devoid of fermentative powers.

Serological characteristics

All freshly isolated strains possess a common antigen, antigen 1, and may also have one or more of antigens 2, 3 and

4; thus fresh isolates can be typed. On subculture the antigenic structure becomes less distinct as isolates lose their pathogenicity.

Epidemiology of whooping cough

Infection is acquired from other *cases* of whooping cough and the fact that fresh cases occur without any known association with typical cases does not imply that a carrier state exists; on the contrary, *Bord. pertussis* cannot be isolated from healthy individuals and infection can arise from cases which are not recognized as suffering from the disease either because the attack is naturally mild or has been modified by active immunization.

Bord. pertussis is most probably spread by direct droplet spray.

Whooping cough can occur at all ages but the incidence is greatest in pre-school children and the fact that a case is most infectious in the catarrhal stage of the illness, and before paroxysms allow a clinical diagnosis, probably explains its epidemic spread. Epidemics occur in regular cycles at two or three-yearly intervals.

Prevention

Since the case is most infectious before the classical symptoms appear, isolation is relatively ineffective in preventing secondary cases if it is delayed until the patient is in the paroxysmal stage of the illness. A high degree of protection is afforded by active immunization as a potent vaccine will reduce the attack rate in home contacts of cases from approximately 90% in the unprotected siblings to less than 5% in contacts who have been fully immunized. Since all deaths from whooping cough occur in the first year of life, immunization should be carried out as early as possible, i.e. between the 2nd and 6th month after birth.

OTHER BORDETELLAE

Bord. parapertussis is morphologically identical with *Bord. pertussis* but grows more rapidly on Bordet-Gengou medium and its colonies are very similar to those of the whooping cough bacillus but the underlying medium is discoloured by a brown pigment.

Parapertussis strains are serologically distinct and although they are occasionally associated with cases of whooping cough the illness is much less severe than that caused by *Bord. pertussis*.

PASTEURELLAE

P. PESTIS

Morphology

Short bacilli with rounded ends, $1.5\mu \times 0.7\mu$. Capsulate in the tissues and on first isolation in the laboratory; non-motile, non-sporing. Bi-polar staining is characteristic and helpful in identification as, when methylene blue is applied, the poles of each bacillus stain more deeply than the central portion.

Cultural requirements

Aerobic and facultatively anaerobic. Temperature range for growth is $14°–42°C$ with an *optimum of 27°C*, although subcultures will grow at $37°C$; grows on ordinary media but growth is enhanced if blood is added to the medium.

Cultural appearances

Colonies are at first small (1 mm in diameter or less), transparent and with a circular edge; on continued incubation they may increase to 3–4 mm in diameter and have an irregular edge.

Biochemical activities

Glucose, mannitol and maltose are fermented without gas production and *P. pestis* will grow on MacConkey's medium in contrast to haemophili, bordetellae and other pasteurellae with the exception of *P. pseudotuberculosis*. However, unlike the plague bacillus, the latter is motile if tested at 22°C.

Serological characteristics

Strains are serologically homogeneous and also share certain antigens with other Pasteurellae.

Animal pathogenicity

Since plague is epizootic in wild rats it would be surprising if the causal organism did not affect rodents used in laboratory diagnostic procedures. If a white rat is injected subcutaneously with a freshly isolated strain of *P. pestis* it dies in 2–3 days; post-mortem examination reveals a necrotic lesion at the point of injection and the regional lymph glands are also involved. In addition to these sites the bacilli can also be recovered from the heart blood and from the enlarged spleen.

Epidemiology of plague

Plague is primarily a disease of rats and other rodents and man is only involved as a secondary host under certain environmental circumstances. In the past, human epidemics were greatly feared and rightly so; for example, in the Black Death of the 14th century, it has been estimated that at least 25% of the population of Europe died.

The bacilli are spread from the rodent host by rat fleas such as *Xenopsylla cheopis*; the flea takes a blood meal from the infected rodent and the bacilli multiply in its stomach and proventriculus, which becomes blocked. If the flea alights on a new host and attempts another blood meal, regurgitation from the blocked proventriculus ensures that plague bacilli are inoculated into the bite wound.

In man there follows an incubation period of 3–7 days before the related lymph glands show evidence of infection; these swell and the surrounding tissues become oedematous so that a primary bubo is formed. In *bubonic* plague secondary buboes may develop along the lymphatic system and if the organisms spread into the blood stream the patient dies from septicaemic plague. Some cases are primarily septicaemic and death may occur before buboes develop fully.

In bubonic plague, infection rarely spreads to other men and fresh cases in an epidemic are individually infected by rat fleas. If, however, pulmonary infection results during a septicaemic phase the patient may act as a source from which other people may acquire *pneumonic plague* either from dust particles contaminated from the sputum or by infected droplets.

Prevention

Rodent control is vitally important, particularly to prevent their importation by a ship or plane whose voyage started from a country where plague occurs. Various vaccines are available but the degree of protection which they give is variable, particularly that made from dead bacilli, e.g. Haffkine's vaccine.

Patients must be treated in isolation and when an epidemic begins intensive flea control should be undertaken as well as rat control. Personnel working in the epidemic area should wear protective clothing which is dusted regularly with insecticide and in addition insect repellant cream should be used. Prophylactic dosing with tetracyclines has been recommended for medical and nursing personnel dealing with cases.

OTHER PASTEURELLAE

Only a few of the other members of the genus *Pasteurella* are pathogenic to man and they are rarely encountered; they do, however, cause infection in many animals and birds. Other species are morphologically indistinguishable from *P. pestis* but can be differentiated from it and from each other by biochemical means and also by their varying virulence to different animals.

P. septica is not only carried by several domestic animals and poultry but can cause disease in the natural host. Wounds in man inflicted by the bites of dogs or cats may be infected with such strains which apparently delay healing of the wound.

P. haemolytica may be carried by healthy cattle and sheep but occasionally causes pneumonia in these animals; non-pathogenic to man.

P. pseudotuberculosis causes a tuberculosis-like disease in several animals including guinea-pigs. It may do this in laboratory guinea-pigs, and so confuse the bacteriologist performing necropsy on animals infected by it which have been injected with pathological material from a suspect human case of tuberculosis.

P. psuedotuberculosis is not acid-fast so that although the lesions it causes may resemble those of tuberculosis, a Z-N stained film of material from them rapidly resolves any doubt as to the nature of the causal organism. *P. pseudotuberculosis* differs from other members of the genus in being motile when grown at 22°C and, like *P. pestis*, growth occurs on MacConkey's medium but growth is more abundant and more rapid than that of the plague bacillus.

BRUCELLAE

BR. ABORTUS

Morphology

Small cocco-bacilli and frequently almost coccal in appearance; approximately $0.4\mu–0.6\mu$. Rarely capsulate; non-motile and non-sporing.

Cultural requirements

Aerobic but requires presence of 5–10% CO_2 for primary isolation. Will grow on ordinary media, but growth is accelerated and enhanced if the medium contains animal protein, e.g. on liver-infusion agar. Temperature range for growth is 20°–40°C. Optimum is 37°C.

Cultural appearances

Growth is slow even under optimal conditions; colonies are small, low-convex and with an entire edge. On continued incubation they increase in size and their original translucence is lost and they become yellowish and eventually brown.

Biochemical activities

No fermentative properties unless the carbohydrate substrates are incorporated in a peptone-free, buffered medium; under these conditions *Br. abortus* can be differentiated from the two other members of the genus since it utilizes glucose and inositol.

More commonly, such differentiation is carried out by checking the ability of strains to grow in the presence of

1 in 25,000 basic fuchsin and 1 in 30,000 thionine; *Br. abortus* is inhibited by thionine.

Serological characteristics

In common with the other members of the genus, *Br. abortus* possesses two antigens, A (abortus) and M (melitensis); however, although antiserum prepared against *Br. abortus* contains both A and M antibodies the latter type can be removed by absorption with a suspension of *Br. melitensis* and the absorbed serum then reacts only with *Br. abortus*. The preparation of such absorbed antisera is possible since M antigen predominates in *Br. melitensis* and its relatively poor content of A antigen does not significantly alter the level of A antibody in the serum during absorption.

Animal pathogenicity

Guinea-pigs are susceptible to experimental infection but are not normally used for diagnostic purposes; other experimental animals are less susceptible.

Epidemiology of Brucellosis

Br. abortus occurs naturally in cattle and is responsible for epidemics of abortion in such animals. *Br. melitensis* is primarily pathogenic for goats and sheep whilst *Br. suis* occurs in pigs. These host specificities are not absolute and man is susceptible to infection by all three species. Infection in man can occur in two ways: *first* by consumption of milk and freshly prepared milk products, such as cheese and butter made from infected milk, and *secondly* the organisms can enter via skin abrasions.

In Britain, *Br. abortus* is the species most commonly encountered in human cases and it is interesting that abortion is very rare in women who suffer brucellosis. It is considered that this is because human placentae do not contain *erythritol,* which abounds in the placental tissues of cows, sheep and goats. This is known to stimulate the growth of brucellae and the subsequent placentitis causes abortion.

Brucellosis also occurs as an occupational infection in farmers, veterinary surgeons and butchers. The organisms enter through skin abrasions when the individual is handling infected discharges or carcases.

Prevention

Pasteurization of milk protects the general population against this source of infection; abortion in cattle has been reduced by widespread vaccination but abortus infection is still endemic and the need for pasteurization before milk is consumed has not declined.

Those following occupations which have a risk of infection must be educated in the handling and disposal of infected excreta, dead foetuses and other infected carcases. There is no vaccine available for human use.

OTHER BRUCELLAE

Br. melitensis and *Br. suis* are morphologically identical with *Br. abortus* and their cultural requirements differ only in that they do not require increased CO_2 in their atmosphere for primary isolation. Their colonial appearances do not differ from those of *Br. abortus* but they are differentiated by sensitivity to dyes; *Br. melitensis* is not inhibited either by basic fuchsin or thionine, whereas *Br. suis* is inhibited by basic fuchsin. Similarly *Br. melitensis* does not produce H_2S and members of the other two species do so; *Br. suis* strains from American sources differ from Danish strains only in their ability to produce H_2S.

CHAPTER 17
Vibrios and Spirilla

VIBRIO CHOLERAE

Morphology

Often termed the comma bacillus on account of its curved shape, approximately $2-3\mu \times 0.5\mu$, Gram-negative, motile by virtue of a single terminal flagellum, non-capsulate, non-sporing. Characteristic shape usually lost on *in vitro* cultivation.

Cultural requirements

Aerobic, temperature range for growth 16°–40°C, optimum 37°C. Sensitive to acid pH, thus media must be adjusted accordingly; optimum pH 8.2.

Cultural appearances

Colonies are not distinctive and are variable; a 'typical' colony is large, 2–3 mm in diameter after overnight incubation at 37°C, with an entire edge and is translucent. Older colonies may develop a yellowish colour.

Biochemical activities

Ferments a variety of sugars without production of gas. Gives a positive *cholera-red reaction*, i.e. when grown in peptone water, cholera vibrios produce indole and nitrites so that when a few drops of H_2SO_4 are added to an overnight culture a pink colour develops due to the production of the red compound nitroso-indole.

Does not cause haemolysis when 1 ml of a 48 hr broth culture is added to an equal volume of a 5% suspension of sheep RBC; this is known as the Greig test.

Serological characteristics

Cholera vibrios are serologically homogeneous in regard to their flagellar and major somatic antigens; however, minor somatic antigens allow the differentiation of three serotypes 'Inaba', 'Ogawa' and 'Hikojima'.

Epidemiology of Cholera

Although an individual recovering from cholera may be a temporary carrier for one or two weeks, chronic carriage beyond this period is exceptional; thus *cases* of the disease are the main source of infection. Man is the only host. The main method of epidemic spread is by water supplies contaminated by cases and the temperature growth range allows the vibrios to multiply in water tanks so that explosive outbreaks are commonly seen; case to case infection via contaminated foodstuffs also occurs and insanitary methods of disposing of excreta may allow insects to contaminate food.

Prevention

In the past cholera occurred pandemically and visited Europe and North America in epidemic waves even in the latter half of the last century; its elimination from many countries was due primarily to the creation of water supply systems which, in addition to incorporating physical and chemical purification techniques, also eliminated the possibility of the water being contaminated by man after purification had taken place. In countries where safe water supplies are not yet available emergency treatment, with hypochlorite, of bulk supplies in storage tanks is practicable and, in any case, water for human consumption or for use in food preparation must be boiled.

Obviously cases of the disease should be treated in isolation and apart from combatting the fluid and electrolyte imbalance caused by the profuse diarrhoea which would otherwise kill the patient, the administration of a suitable antibiotic will eliminate the vibrios and shorten the period in which the convalescent is a risk to the community.

Cholera vaccine given in two doses at 7-day intervals may give some degree of protection for a month or two but its real value is not known.

OTHER VIBRIOS

The El Tor vibrio has many similarities to *V. cholerae* and infections caused by El Tor vibrios are clinically indistinguishable from those caused by the classical strains.

It can be distinguished from *V. cholerae* in giving a positive Greig test, by its ability to cause haemagglutination of fowl RBC and by its resistance to polymyxin B. In addition, El Tor strains are resistant to the activity of phages which lyse *V. cholerae*.

Many other vibrios can be identified and several are causally related to diseases in various animals; others are saprophytic in water.

SPIRILLUM MINUS

Morphology

$2-5\mu \times 0.5\mu$, rigid spiralled organism; Gram-negative and actively motile with bipolar lophotrichous flagella. Non-capsulate and non-sporing.

This organism, which is the only member of the genus *Spirillum* known to be pathogenic to man has not been successfully cultured *in vitro* and hence its other biological characteristics are unknown.

Sp. minus is a natural parasite of rats, mice and other rodents and man may suffer infection if bitten by an infected rodent. In cases of rat-bite fever the organism can be demonstrated in material from the bite wound, the regional lymph glands and on occasion in the blood either by microscopy or by intraperitoneal inoculation of material into a guinea-pig. The animal develops a chronic febrile illness and spirilla can be detected in its blood and also in the lymph glands 7–14 days after inoculation; at post-mortem the organism can be demonstrated in most tissues and organs.

Cases of rat-bite fever are sporadic and preventive measures are obvious.

CHAPTER 18

Spirochaetes

1 BORRELIAE

BORR. VINCENTII

Morphology

5–20μ × 0.4μ, with 3–8 coils which are irregular in amplitude; Gram-negative, non-capsulate, non-sporing.

Cultural requirements

Strictly anaerobic and very difficult to grow in pure culture; cultivation is never attempted for diagnostic purposes.

Borr. vincentii is found in *small* numbers in the healthy human mouth and in association with a large, cigar-shaped, bacillus *Fusobacterium fusiforme*. In mouths in a poor state of dental hygiene and particularly when the individual is suffering other buccal infection or is malnourished, e.g. suffering from Vitamin C deficiency, very much larger numbers of these two organisms can be detected, particularly in films made from the lesions of Vincent's infection. There is doubt as to whether the symbionts are primarily responsible for the infection but since it responds to penicillin quite dramatically there is no doubt that they have a causal rôle, perhaps as secondary pathogens.

The disease is, therefore, endogenous and usually sporadic, but occasional epidemics have been reported; prevention depends on maintenance of oral hygiene and correction of any nutritional deficiencies.

OTHER BORRELIAE

Other members of the genus cause diseases in domestic animals and poultry and some occur as commensals of man; however, one or two are incriminated as the causal agents of relapsing fever and are transmitted from case to case by lice or ticks.

Borr. recurrentis, the causal organism of European relapsing fever, is similar in morphological appearance to *Borr. vincentii* except that the coils have a regular amplitude. It can be harvested in a variety of fluid media under anaerobic conditions but diagnosis usually involves only the demonstration of the organism in films of peripheral blood or by injecting white rats with a blood sample from the patient. The organism is transmitted from case to case by the body louse which infects a bite-wound with its excreta; alternatively, lice may be crushed when the individual scratches, thus inoculating himself through skin abrasions.

Borr. duttonii is identical with *Borr. recurrentis* with the exception of its distinctive antigenic make-up; it causes West African relapsing fever and is tick-borne. Spread of infection may be from case to case via the tick or may be transmitted by the tick to man from primary hosts, e.g. rodents and pigs. A tick may remain infected for some years after having a blood meal and spirochaetes are also transmitted transovarially to consecutive generations of ticks.

Several other borreliae have been incriminated as causing relapsing fever in other countries and on insufficient evidence have been given species designation.

2 TREPONEMATA

TR. PALLIDUM

Morphology

$6-14\mu \times 0.1\mu$, and possessing 6–12 regular coils. So fine is the spiral filament that it cannot be seen in film preparations stained by ordinary methods; the organisms can be seen by dark ground illumination or, alternatively, by the ordinary light microscope after their size has been increased by silver impregnation staining.

Tr. pallidum has not been grown *in vitro* and hence we know nothing of its biochemical or other characteristics; nevertheless, in syphilitic patients antibodies can be detected which immobilize *Tr. pallidum* strains maintained in the laboratory by intratesticular inoculation of rabbits. Such antibodies can also fix complement but in these tests the stock antigen is an extract from normal tissues and does not contain *Tr. pallidum*.

K

Epidemiology of Syphilis

Sources are cases in the primary and secondary stages of the illness and with few exceptions infection is contracted during sexual intercourse; exceptions to this rule include acquisition of a primary lesion on the hand of the unsuspecting medical practitioner. *Tr. pallidum* dies so rapidly outside the host tissues that transfer by indirect contact is extremely rare.

Congenital syphilis, i.e. intra-uterine infection of the foetus via the placenta, is very much less common nowadays and this reduction in incidence reflects the dramatic advances in antenatal care which have occurred in the last 30 years and also the efficiency of antibiotics in treating adult cases.

Prevention

Since a single source gives rise to several new cases, sometimes as many as 30 or more, prevention demands speedy diagnosis and treatment to render the patient non-infectious as rapidly as possible and, simultaneously, the tracing of contacts of each new case must be rigorous. Whilst in some countries compulsory powers are used to trace contacts, prevent default from treatment, etc., in Britain the voluntary system of control is preferred in combination with health and sex education.

OTHER TREPONEMATA

Many organisms resembling *Tr. pallidum* occur commensally in man and some of these are resident on the genital mucosa and might confuse the diagnosis when a suspected case of primary syphilis is being investigated; however, if care is taken to thoroughly clean the area before collecting exudate from the lesion, commensals such as *Tr. gracile* will not be collected in the specimen.

Although *Tr. pallidum* is the only pathogenic member of the genus found in Britain, other treponematoses occur but almost entirely in tropical countries. None of these is sexually acquired and infection is probably transmitted by personal contact or by drinking from contaminated communal cups as is the case with *bejel*, which is essentially a disease of children, and which presents with lesions similar to those of secondary syphilis without a primary lesion having occurred.

Another non-venereal infection which is better known than bejel is *yaws*, where transmission of the causal organism, *Tr. pertenue*, is by direct contact or by flies which feed on lesions. Organisms which survive in the proventriculus for several hours are regurgitated when the fly has another feed. *Tr. pertenue* is biologically identical with *Tr. pallidum*.

Tr. carateum is responsible for the disease *pinta* which occurs particularly in South American countries; this organism is indistinguishable from that causing syphilis but it is likely that they differ in antigenic constitution since individuals suffering from syphilis can acquire pinta and vice versa.

3 LEPTOSPIRAE

L. ICTEROHAEMORRHAGIAE

Morphology

6–$20\mu \times 0.1\mu$, coils are very numerous, small and close together and although visible with dark-ground illumination they are obscured if stained preparations are examined; one or both ends of the organism are recurved on the body.

Cultural requirements

Unlike members of the genera *Borrelia* and *Treponema*, leptospirae can be isolated *in vitro* for diagnostic purposes and although solid media are available for research purposes the organism grows more readily in fluid media, e.g. Stuart's, provided these contain animal serum. Essentially microaerophilic and with an optimum temperature of 30°C. Sensitive to acid environments.

Biochemical activities

No knowledge available.

Serological characteristics

Certain antigens are shared with some other pathogenic leptospirae, but antigens specific for *L. icterohaemorrhagiae* allow us to prepare antiserum which can be absorbed by leptospirae possessing only the shared antigens. Such absorbed antiserum can then be employed in agglutination tests to identify newly isolated strains. Because of the morphological similarity of all leptospirae and our inability to use cultural or biochemical methods of establishing the separate identities of various strains, we rely entirely on serological methods of identification.

Animal pathogenicity

Young guinea-pigs and golden hamsters are most susceptible and after intraperitoneal inoculation with pathological material containing leptospirae, e.g. patient's blood, or with freshly isolated laboratory cultures, the animal shows progressive fever and then jaundice and dies in 10–14 days. At post-mortem there is generalized jaundice and haemorrhage, the spleen and adrenals are enlarged and friable and the spirochaetes can be recovered from many organs and are most numerous in the liver.

Epidemiology of Weil's disease

Weil's disease is the name given to human infection caused by *L. icterohaemorrhagiae*. This organism lives normally in the brown rat and there leads an essentially commensal existence and is localized in the rat's kidneys and is thus discharged in the urine. Organisms discharged into man's environment may survive for only an hour or two unless they are in moist surroundings with an alkaline pH; thus, in Britain, Weil's disease is to a large extent associated with certain occupations such as mining, fish-gutting, etc., where people are working in damp, rat-infested places.

The organisms penetrate through cuts and abrasions in skin and mucous membranes and occasional cases arise through people bathing in canals, rivers or stagnant ponds which are populated by rats. Like almost all other zoonoses, Weil's disease does not spread from patients to other human beings.

Prevention

Individuals engaged in high-risk occupations must be taught about sources of infection and the methods of its spread, and wherever possible should wear protective clothing; eradication of rats, and rodent proofing, should be undertaken.

Washing down of working surfaces, e.g. slabs in fish-gutting halls, with acid solutions also assists protection of workers.

OTHER PATHOGENIC LEPTOSPIRAE

Numerous serotypes have been identified throughout the world and these are frequently associated with various natural hosts; apart from *L. icterohaemorrhagiae* only one other type, *L. canicola*, occurs in Britain. The resultant illness in man is Canicola fever which is much milder than Weil's disease, frequently anicteric and rarely fatal. Dogs and pigs act as natural hosts and man becomes infected by mopping up dog's urine or by working in piggeries.

CHAPTER 19

Higher Bacteria

Within the higher bacteria two families, *Streptomycetaceae* and *Actinomycetaceae*, have medical significance. In the former family some members of the genus *Streptomyces* produce clinically useful antibiotics, such as streptomycin from *Streptomyces griseus*.

Higher bacteria pathogenic to man and animals belong to the *Actinomycetaceae* and here two genera, *Actinomyces* and *Nocardia* must be considered.

1 ACTINOMYCES

ACTINO. ISRAELII

Morphology

Branching filaments approximately 1μ in diameter which interlace to form a mycelium. Gram-positive, non-motile, non-capsulate, non-sporing. Fragmentation of the filaments gives rise to bacillary and coccal forms. In sections of actinomycotic lesions stained by Gram's method, club-shaped structures which stain Gram-negatively can be noted at the periphery of the Gram-positive mycelium. A similar section stained by Ziehl-Neelsen's technique shows that these peripheral clubs are acid-fast if 1% H_2SO_4 is used in attempted decolourization instead of a 20% solution as for *M. tuberculosis*.

Cultural requirements

Anaerobic or microaerophilic atmosphere is required; optimum temperature for growth is $37°C$ and growth does not take place at temperatures much above or below the optimum; grows on ordinary agar but greatly enhanced if serum or blood is incorporated.

Cultural appearances

Colonies show considerable variation but are nodular, opaque and ochre in colour; firmly adherent to underlying medium.

Biochemical activities

These are saccharolytic but there is much variation among strains and thus no reliance can be placed on such reactions.

Serological characteristics

Antigenic analysis permits the recognition of four serotypes of *Actino. israelii* but at present such serotyping is essentially of academic importance.

Animal pathogenicity

Experimental inoculation of rabbits and guinea-pigs may or may not result in nodular granulomatous lesions hence such methods are not used in diagnosis.

Epidemiology of Actinomycosis

Infection is endogenous from organisms which normally fulfil a commensal rôle in the mouth. Almost three-quarters of all lesions occur in the oro-facial region and the disease is apparently precipitated by trauma since there is frequently a history of, for example, dental extraction or, in the case of abdominal actinomycosis, recent appendectomy or external trauma.

In Britain actinomycosis is most prevalent among male agricultural workers and it was once thought that the human disease was acquired from cattle whose mouths in health often contain *Actino. bovis*. As well as this commensal existence, this organism can produce a disease, lumpy jaw, the bovine counterpart of human actinomycosis. In addition to its high incidence in agricultural workers actinomycosis is much more common in men than in women (4 : 1) and the majority of cases occur in the age group 10–29 years.

OTHER ACTINOMYCES

Actino. bovis is morphologically identical with *Actino. israelii* but it can grow aerobically. Colonies are smoother than those of *Actino. israelii* and do not adhere to the medium; biochemical activity is variable from strain to strain but unlike *Actino. israelii*, *Actino. bovis* strains hydrolyse starch. Serologically unrelated to *Actino. israelii types*.

2 NOCARDIA

NOC. MADURAE

Morphology

Basically similar to *Actino. israelii* but filaments are narrower, i.e. 0.5μ.

Cultural requirements

Obligate aerobe with an optimum growth temperature of 37°C. Grows on a wide range of media.

Cultural appearances

Like *Actino. israelii*, colonies are very adherent to the surface of the medium. On agar, colonies are at first small, round and convex but later they increase in size and become opaque with a rosette appearance. They are difficult to emulsify.

Biochemical activities

Has no saccharolytic powers.

Serological characteristics

Knowledge of antigenic structure is incomplete.

Noc. madurae is one of the causes of madura foot, a granulomatous infection confined to the feet which occurs only in the tropics and some sub-tropical countries. In addition to other nocardiae some cases of madura foot are caused by true fungi, e.g. *Madurella*.

There are more than 40 other members of the genus *Nocardia* but the majority of these are saprophytic; however, at least one other species, *Nocardia asteroides*, is pathogenic for man and in Britain this organism is responsible for pulmonary nocardial infection. Morphologically similar to *Noc. madurae*, it is also *acid-fast*, a strict aerobe and its colonies are *star-shaped*. It produces acid from glucose.

CHAPTER 20
Specimens: Collection, delivery and processing

This section gives a synopsis of the methods of collection, delivery and the laboratory processing of the more commonly encountered specimens; it is not comprehensive.

Unsatisfactory specimens are still submitted to laboratories but perhaps less frequently than in the past; this improvement reflects a greater awareness on the part of the clinician of the very obvious fact that no matter how sophisticated modern laboratory procedures may be, they cannot compensate for certain basic errors in the collection and transmission of specimens.

Special requirements for the various types of specimen are given in this chapter, but firstly some general points must be emphasized which, although glaringly obvious, are ignored from time to time.

1 All specimens must be accompanied with details of the patient's name, sex, age and address and the nature of

the specimen; these are *minimal* requirements solely for the purposes of identification. Similarly, the doctor who submits the specimen should give his identity and address; in every laboratory there is a 'lost lambs' file containing request forms which may be completely blank or with insufficient information to allow identification of the patient and/or doctor. Each of these forms is not only a source of frustration to the laboratory staff but on occasion the patient's health or even life may be endangered and sometimes the community is placed at unnecessary risk, e.g. a specimen of faeces from an unknown source may be found to contain typhoid bacilli.

Additional information which should accompany the specimen includes the clinical diagnosis, duration of the illness, information regarding the patient's immune state, e.g. if previously immunized with TAB when serum is submitted for Widal testing, and whether antimicrobial therapy has been given.

2   No antiseptics or other antimicrobial agents should come into contact with the specimen; all specimen containers should be sterile before use and they can be supplied in this state by the laboratory.

3   In collecting the specimen care must be taken to avoid contamination of the outside of the container otherwise the clinician, the person delivering the specimen to the laboratory, and the laboratory staff may be infected.

4   Transmission of the specimen to the laboratory should be within a few hours; limits for certain types of specimen are specified under the individual headings.

5   If specimens are delivered by postal service certain regulations must be observed and these are known to the investigating laboratory which can provide suitable containers for this purpose. The important points in the regulations are:

(a) the specimen container should be sealed in such a way as to prevent the escape of pathological material and it must be placed in a suitable case which is padded with absorbent material;

(b) suitable cases are prescribed and are made either from wood or leatherboard;

(c) the outer envelope must be clearly marked 'Pathological Specimen' and 'Fragile with Care'; specimens can only be sent by letter post and not by parcel post.

BLOOD CULTURE

Certain basic facts must be stressed regarding the collection of venous blood for the attempted isolation of pathogens. Firstly as venepuncture is required it is essential that a fully aseptic technique is followed so that *the patient is not infected as a result of the diagnostic procedure*. Similarly extraneous contamination of the specimen must be avoided otherwise the laboratory may issue a misleading report. Since the bacterial population of the blood in bacteraemia may be very small, e.g. only one or two organisms per ml, at least 5 ml of blood should be collected. The laboratory can supply a variety of media for blood culture and each has its advantage in isolating particular pathogens so that the clinician who suspects that a patient is suffering from a particular infection should consult the bacteriologist concerning which media are most suited to the particular case.

It is important that the bacteriologist should be informed of any antibacterial therapy which the patient is receiving since he can then incorporate in the medium certain enzymes which will eliminate the continued activity of the antibacterial agent in the blood once it has been introduced to the culture medium. For example, penicillinase can be added to the broth if the patient has been receiving penicillin and this enhances the prospects of isolating any organisms present in the specimen; of course since only 5 ml of blood are added to 50 ml of broth any antibiotic present will be significantly diluted.

Delivery to the laboratory

Once inoculated, the blood culture bottle must be returned to the laboratory immediately and if this is not possible it must be maintained at a temperature as near as possible to, but not exceeding, $37^{\circ}C$.

Laboratory procedure

After 18–24 hrs incubation at $37^{\circ}C$ the blood-broth mixture is sampled by removing aseptically two or three loopfulls which are plated on to two blood agar plates, one of which is incubated at $37^{\circ}C$ under aerobic conditions whilst the other is incubated anaerobically. Such sampling is repeated every 48 hrs for at least 2–3 weeks before a negative report is issued; any organisms which are isolated are of course subjected to complete identification.

When the patient is suspected to be suffering from brucellosis a special blood culture bottle is used. This has a layer of liver-infusion agar lying along one side (in addition to the broth) and after delivery to the laboratory CO_2 is introduced to give a 5–10% concentration and the bottle is then incubated. The agar surface is inoculated by carefully tilting the bottle so that the blood-broth mixture flows gently over the medium; thus the bottle need not be opened every 48 hrs for sampling and any brucellae present, particularly *Br. abortus*, will have optimal conditions for growth. Although colonies will usually be visible within 7 days the blood culture bottle should be incubated for 4 weeks before a negative report is issued in those instances when there is no growth.

CLOT CULTURE

In suspect cases of enteric fever the chances of isolating *S. typhi* or one of the paratyphoid bacilli during the bacteraemic phase of the illness are enhanced if clot culture is used instead of blood culture, even if a bile-salt broth is used in the latter method.

5 ml of venous blood are obtained by venepuncture and transferred into a suitable sterile container, e.g. a 2 oz screw-capped bottle, and allowed to clot. The serum is then removed aseptically and the clot is lyzed by adding to it 15 ml of 0.5% bile-salt broth containing 1,500 units of streptokinase. Thereafter the mixture is incubated at 37° and examined every 48 hrs in the same way as for blood culture. Not only is the isolation of salmonellae more frequently made by clot culture but there are also additional advantages; the clinician does not require a supply of special blood culture bottles and a Widal test can be carried out on the separated serum so that the titre obtained at this stage can be used as a base-line for the results of Widal tests later in the illness.

CEREBROSPINAL FLUID

The comments regarding the patient's safety and the prevention of contamination of the specimen made under 'blood culture' are equally pertinent in performing a lumbar puncture to obtain a specimen of cerebrospinal fluid (CSF).

Delivery to the laboratory

Delivery must be very rapid and the fluid should be warm when it arrives; if this is impossible then the container must be held at 37°C and in any case *not more than 4 hrs should elapse* before laboratory processing is undertaken. This time and temperature requirement is dictated by the feeble powers of survival of some of the organisms which may cause meningitis, in particular the meningococcus and *H. influenzae*.

Laboratory procedure

1 *Macroscopic examination* of the specimen should be made for colour, turbidity and for the presence of any deposit or clot; then 1 ml of CSF is transferred aseptically to a tube containing a similar volume of 0.2% glucose broth which is then incubated for 24 hrs at 37°C. The broth enhances the growth of meningococci and pneumococci if either of these species is present and subculture of the CSF-broth mixture, after incubation, is made to two blood agar plates, one of which is incubated aerobically and the other in an atmosphere containing 5% CO_2.

2 *A white cell count* is performed and the remaining CSF is then centrifuged at 3000 rev/min for 5 mins; the supernatant is removed aseptically and sent for biochemical examination.

3 *The centrifuged deposit* is then used to make films which are stained by Gram's method and the deposit is also used to inoculate two blood agar plates which are incubated under the same conditions as the subcultures from the CSF-broth mixture as above.

4 CSF from *cases of tuberculous meningitis* frequently contain a 'spider-web' clot if the container is allowed to stand upright without being disturbed for 1–2 hrs at 37°C and if a film is made with part of the clot and stained by the Ziehl-Neelsen method, tubercle bacilli may be seen.

In addition to microscopic examination of the clot, or alternatively if no clot is present, the CSF is centrifuged and the deposit is used to inoculate tubes of Lowenstein-Jensen medium and is also injected into a pair of guinea-pigs. Animal inoculation yields a higher proportion of positive results than does cultivation of the same specimen.

URINE

Although *mid-stream specimens* of urine from male patients have always been accepted as satisfactory for bacteriological examination, it is only in the last few years that there has been general agreement that such specimens from female patients are also reliable; in both sexes, provided that the patients are properly instructed, contamination of the urine can be avoided and the bacteriological results obtained with mid-stream specimens are as reliable as those with catheter specimens of urine.

Catheterization must be avoided wherever possible since *the procedure carries a definite risk of introducing infection.*

In the male the foreskin is retracted, the glans washed with soap and water and after voiding a little urine to flush commensal organisms, e.g. staphylococci and diphtheroids, from the anterior urethra the mid-flow of urine is collected in a suitable sterile screw-capped container; in women the ano-genital region is thoroughly washed and then with the labia widely separated the mid-stream urine is collected in a sterile wide-mouthed container, e.g. a 12 oz honey-pot.

Delivery to the laboratory

Since urine is an excellent culture medium the specimen must be transported to the laboratory within 1–2 hrs otherwise any organisms present in the specimen will multiply and the results of quantitative examination will be misleading; if this time limit cannot be met the specimen must be stored in a refrigerator until collected or alternatively it may be delivered in a special container which maintains a low temperature.

Laboratory procedure

1   *Microscopic examination.* 5 ml of the specimen is centrifuged at 3000 rev/min for 5 mins and the deposit resuspended in a small amount of the supernatant after the larger part of the latter has been decanted. A wet film is made from the resuspended deposit and examined for the presence of bacteria, pus cells and red blood cells.

2   *Cultivation.* The resuspended deposit is also used to inoculate a blood agar plate and a plate of MacConkey's medium and if on microscopic examination any bacteria are present a primary sensitivity plate should also be seeded with the deposit. These plates are incubated overnight at 37°C and species identification is then made and, if necessary, subculture sensitivity tests are performed.

3   *Bacterial count.* Provided that the specimen is properly collected and is delivered to the laboratory within the conditions mentioned above then an estimation of the number of bacteria per ml of urine is valuable in deciding whether the bacterial population of the specimen is from established infection or has resulted merely from contamination. Separate dilutions of urine, 1 in 100 and 1 in 1000, are made in nutrient broth and then a known volume of each, e.g. 0.1 ml, is spread over the surface of two nutrient agar plates which are then incubated for 18–24 hrs at 37°C. The colonies are then counted and assuming that each colony represents one organism in the original inoculum the number of bacteria per ml can be estimated.

Counts of 10^3 organisms or less per ml almost invariably reflect a contaminated specimen whereas if 10^5 or more organisms are found to be present in 1 ml this is indicative of infection. Counts in the region of 10^4 bacteria per ml may result either from contamination or infection and in such cases the count should be repeated on a fresh specimen; such counts probably result from contamination if more than one bacterial species is present.

Urinary tuberculosis

In cases suspected of suffering from tuberculous infection of the urinary tract the specimen submitted to the laboratory should be either a 24 hr collection in a suitable container, e.g. a Winchester bottle, or alternatively three consecutive early morning specimens which are pooled in the laboratory before examination.

L

Contamination of such specimens with commensal acid-fast *M. smegmatis* can be reduced by thorough washing of the genitalia before urine is passed.

Before such specimens can be examined they must be processed to reduce their volume, to kill organisms other than tubercle bacilli and to liquefy mucus or cells within which tubercle bacilli may be trapped. This processing is known as concentration and several methods are available but all make use of the relative resistance of tubercle bacilli to acids, alkalis or other mucolytic agents.

After concentration a film is made and stained by the Ziehl-Neelsen method and the concentrate is also used to inoculate slopes of Lowenstein-Jensen's medium and two guinea-pigs.

THROAT SWABS

It is essential that the throat should be examined and swabbed under adequate illumination; a spatula must be used to depress the tongue and the use of the handle of a domestic spoon or fork in place of a spatula must be condemned. Such narrow 'spatulae' only serve to force the edges of the tongue upwards and thus obscure a view of the tonsillar areas; if antiseptic lozenges have been sucked or if the patient has gargled then swabbing should be deferred for at least six hours.

Delivery to the laboratory.

Swabs should be processed in the laboratory within 12 hrs of being taken; some types of cotton-wool have a lethal effect on certain organisms such as *Strept. pyogenes* and if the delay between taking the swab and delivery to the laboratory is likely to be more than the limit stated then serum-coated swabs should be used. These prolong the survival time of bacteria on the swab.

Laboratory procedure

Bacteria which may be responsible for sore throat are *Strept. pyogenes*, Vincent's organisms and *C. diphtheriae*; thus all throat swabs should be inoculated on to the relevant diagnostic media.

1 *Strept. pyogenes*. Two plates of crystal violet blood agar (CVBA) are inoculated, a bacitracin disk is placed in the well-inoculum area of each, and one plate is inoculated aerobically and the other under anaerobic conditions. The incorporation of a 1 in 500,000 concentration of crystal violet reduces the growth of commensal organisms; anaerobic conditions enhance the growth of *Strept. pyogenes* and if β-haemolytic streptococci grow on the plate but are inhibited in the vicinity of the bacitracin disk the strain very probably belongs to group A, i.e. *Strept. pyogenes* (Plate 12, facing p. 57).

2 *C. diphtheriae*. A tube of Loeffler's inspissated serum medium and a plate of tellurite medium are inoculated; the former gives very rapid growth from which films are made and stained by Gram's method and also to detect the presence of volutin granules, e.g. by Albert's method. Any growth showing Gram-positive bacilli which also possess volutin granules demands that the material from the Loeffler slope should be subinoculated to blood agar for further identification.

Growth of diphtheria bacilli is more slow on tellurite-containing media and films of colonies from such media do not always show volutin granules in the bacilli. Colonial morphology on tellurite media allows differentiation of the three biotypes of *C. diphtheriae* from each other and from commensal diphtheroid species.

All colonies suspected to be those of diphtheria bacilli must be examined for fermentative ability and then be used for tests demonstrating the production of diphtheria toxin.

3 *Vincent's organisms*. These organisms cannot be cultured for diagnostic purposes; hence the diagnosis of Vin-

cent's infection depends on the microscopic demonstration of *large numbers* of *Borr. vincentii* and fusiform bacilli. Films are usually stained with dilute carbol fuchsin for 10 mins.

In no other instance are films, made directly from the throat swab, of any diagnostic significance since the commensal flora of the throat and mouth are morphologically similar to and in some instances identical with the commonly encountered pathogens.

Crystal violet and tellurite are inhibitory substances and thus the swab should be inoculated on to Loeffler's media, CVBA and the tellurite medium in that order; then the film for Vincent's organisms is made.

SPUTUM

Although there are occasions when material from the lower respiratory tract can be obtained by bronchoscopy, most specimens comprise sputum coughed up from the bronchi. Sputum is therefore mixed with organisms in the throat and mouth before expectoration and many species may be present but in varying proportions.

It is important that the patient should be instructed that the specimen must be *coughed up* and that material hawked from the post-nasal space or saliva are valueless.

In young children who cannot expectorate or in adults who are not producing sputum a laryngeal swab may be taken.

Delivery to the laboratory

Many of the species involved in chest infections have poor powers of survival outside the body, so that specimens must be processed in the laboratory within a few hours of being collected.

Laboratory procedure

Homogenization of the specimen must precede cultivation since by sampling different parts of the raw specimen different organisms are found; thus cultures made from homogenized sputum are much more representative of the total bacterial flora.

Homogenization can be effected either by adding sterile glass beads and vigorously shaking the specimen for 15–30 mins, or by the more efficient but more time-consuming pancreatin digestion technique. To the specimen is added an equal volume of 1% buffered pancreatin and the mixture is thoroughly shaken, then placed in a 37°C water bath for 1 hr during which time the mixture is thoroughly shaken every 15 mins.

Thereafter a film is prepared and stained by Gram's method and two plates of blood agar medium are inoculated; one plate is incubated aerobically and the other in an atmosphere with a 5% concentration of CO_2. This latter plate assists the growth of *H. influenzae* and pneumococci; on both plates an 'Optochin' disk can be placed in the well inoculum area to give rapid differentiation of pneumococci from *Strept. viridans* and if a penicillin disk is also placed in the well inoculum it will inhibit many organisms in its immediate vicinity and give a selective zone in which any *H. influenzae* can grow, often in pure culture.

In cases of suspected pulmonary tuberculosis, at least six specimens of sputum should be examined. These are concentrated and the concentrate used to make films for staining by Ziehl-Neelsen's method and also to inoculate slopes of Lowenstein-Jensen's medium and guinea-pigs. In Britain nowadays many cases of pulmonary tuberculosis are detected at a much earlier stage in the illness and the tubercle bacilli are often much scantier than in established or chronic tuberculosis. Cultures should therefore be incubated for at least eight weeks before they are judged to be negative.

FAECES

Specimens of faeces are preferable to rectal swabs in the
bacteriological investigation of cases of intestinal infection.
A disadvantage of rectal swabs is that they are frequently
not taken correctly and are only anal swabs; *it is essential*
that the swab should pass into the rectum. Certain pathogenic species, including *Sh. sonnei* which is the commonest
intestinal pathogen in Britain, do not survive for very long
on plain cotton-wool swabs and therefore rectal swabs
should be serum-coated; the use of rectal swabs denies
macroscopic examination of the faeces and because of the
small inoculum available for plating out etc, *microscopic*
examination for the ova of protozoal parasites can be rarely
undertaken.

Delivery to the laboratory

Specimens of faeces should be examined within a few hours
of being collected and if the delay between collection and
processing in the laboratory is likely to exceed 18–24 hrs
then the specimen should be thoroughly mixed with an
equal volume of buffered glycerol-saline. Rectal swabs
should be processed immediately they have been taken,
hence their use should be restricted to the investigation of
institutional outbreaks of intestinal infection when the
swabs can be delivered to the laboratory within a very
short time and before they have dried.

Laboratory procedure

Specimens of faeces are examined naked-eye for the
presence of blood (fresh or altered), mucus and for any
worms which may be present.

Before any further examination is undertaken, any specimen which is solid or formed must be mixed with a quantity

of sterile physiological saline sufficient to give a thick emulsion.

Microscopic examination

Gram-stained films of faeces are of no diagnostic value because of the morphological similarities between pathogens and the commensal coliforms in the gut. However, microscopic examination of wet film preparations will reveal the presence of pus cells and red blood cells as well as parasites and/or their ova.

Cultivation

Primarily the search is for shigellae and salmonellae which account for most cases of bacterial infection of the intestinal tract, thus plates of deoxycholate citrate agar (DCA) and MacConkey's medium are inoculated from the specimen. In addition to these selective media a sample of faeces should also be emulsified in tubes of two enrichment media, i.e. tetrathionate broth and selenite F broth; after these fluid enrichment media have been incubated for 12–18 hrs at 37°C they are subinoculated to fresh DCA plates.

Any lactose non-fermenting (pale) colonies which appear on either DCA plate or on MacConkey's medium must be investigated for motility and fermentative reactions, and if such colonies give reactions typical for salmonellae or shigellae they must then be identified fully by serological methods.

Some laboratories also carry out routine procedures for the identification of enteropathogenic strains of *Esch. coli* whereas others do so only when requested; such strains grow poorly if at all on DCA but can be harvested from MacConkey's medium. Specimens suspected of containing enteropathogenic *Esch. coli* should also be plated out consecutively on two blood agar plates since some strains do not grow on MacConkey's medium.

There are no colonial or biochemical differences between commensal and enteropathogenic strains of *Esch. coli* so that the latter must be identified by serological methods. At least 10 colonies therefore must be tested by slide agglutination against a polyvalent serum prepared against the six enteropathogenic types and if any reaction is observed the colony is then checked with each of the individual specific antisera to allow type identification.

PUS AND WOUND DISCHARGES

Whenever possible, pus or excised tissue should be submitted for examination. The arrival of a single swab with a light smearing of dried pus accompanied by a request that the bacteriologist should search for pyogenic and anaerobic organisms causes much frustration and is unrealistic. Such swabs dry out very rapidly and many pathogenic species do not survive long, thus, *if swabs must be submitted*, they should be serum-coated and *at least three swabs from the one site* are required if the bacteriologist is to conduct a full investigation.

Dressings from discharging wounds or sinuses are acceptable specimens and should be transported in wide-mouthed, sterile, screw-capped containers.

Delivery to the laboratory

This should be as rapid as possible and if swabs or dressings are submitted they should be processed in the laboratory within 6–12 hrs.

Laboratory procedure

1 *Macroscopic examination* should never be ignored; pus from streptococcal lesions is thin and serous in comparison with that from staphylococcal infection, which frequently

has a gelatinous appearance. In some instances pus resulting from infection with *Ps. pyocyanea* has a distinctive
blue-green colour. In cases of actinomycosis sulphur granules may be noted; these may be semi-transparent or white
if the lesion is early and the characteristic sulphur-yellow
granules are seen in most chronic cases.

2 *Microscopic examination.* Films stained by Gram's
method must always be examined; similarly a Ziehl-
Neelsen-stained film should be prepared if the pus is from a
suspected tuberculosis infection. If sulphur granules are seen
then one should be crushed between two microscope slides,
one slide then being stained by Gram's method and the
other by Ziehl-Neelsen's method but substituting 1% H_2SO_4
for the 20% strength normally used.

3 *Cultivation.* All specimens must be cultured for pyogenic organisms by inoculating a blood agar plate and a
plate of MacConkey's medium; thus the commonly encountered pyogenic organisms, i.e. *Staph. pyogenes* and
Strept. pyogenes are readily isolated, as well as Gram-negative bacilli. A primary sensitivity plate should also be prepared.

In addition, either at the request of the clinician or because of the bacteriologist's suspicions (or intuition), a
search may be made for anaerobic species. One must re-
iterate, however, that the recovery of a member of the genus
Clostridium from pathological material is not synonymous
with clostridial infection and the diagnosis in all such cases
rests solely with the clinician.

The attempted isolation of clostridium is a more sophisticated exercise, *and all media must be incubated under
strictly anaerobic conditions.* A blood agar plate is inoculated and it is recommended that the agar content should be
increased to 6% to discourage spreading of the clostridial
colonies. A half-antitoxin plate may also be inoculated
with the specimen. Several tubes of cooked-meat broth
should be inoculated and then heated at $100°C$ for 5, 10 and
15 mins before being incubated. Such heating destroys non-

sporing pyogenic species so that they cannot contaminate either blood agar or half-antitoxin plates which are inoculated from the cooked-meat broths daily for one week. The latter are incubated throughout this period so that early subcultures will yield the rapidly growing clostridia and later subcultures cater for the more slowly growing members of the genus.

Solid media inoculated directly from the specimen or as subcultures from the cooked-meat broth must be incubated for at least 48 hrs before the absence of growth allows a negative report to be issued.

CHAPTER 21

Antimicrobial Sensitivity Tests

In most people's minds the introduction of antimicrobial agents is equated with Domagk's discovery of the sulphonamides or even with the later discovery of the therapeutic value of penicillin which followed ten years after Fleming's original observations in the laboratory. However, antimicrobial agents have a much longer history even if one ignores the environmental uses of disinfectants and antiseptics.

Although the ancient Egyptians applied many medicaments topically the systemic administration of mercurial preparations in the attempted cure of syphilis was probably the first endeavour to eradicate deep-seated bacterial infection with an antagonistic agent.

Modern antimicrobial therapy was founded by Ehrlich, who studied the effects of dyestuffs on trypanosomes and showed that the latter could develop resistance to dyes which originally had been lethal. Ehrlich is best remembered, however, for his researches with salvarsan and its use in the treatment of syphilis. Ehrlich spoke of 'magic bullets' in the shape of antibodies or of drugs which could kill microorganisms in the patient's tissues but modern antimicrobial agents are more closely analogous to two-edged swords; no one could deny the benefits of chemotherapeutic or antibiotic drugs but at the same time the dangers associated with their abuse, and even their correct use, are not as widely appreciated.

The hazards of antimicrobial therapy can be classified into those affecting the individual patient and those affecting the community.

With regard to the treatment of a patient it is obvious that unless he is suffering from an infection caused by a microorganism that is sensitive to an antimicrobial agent there is no virtue in initiating such specific therapy. It has been reported that probably not more than 10% of the world's production of antibiotics is put to proper use. One can appreciate the pressures on the practitioner to prescribe antibacterial agents not only by pharmaceutical houses but by the patient or his parents or other relatives but such pressures cannot explain the use of these drugs for disorders like headache, toothache, sprained back etc, where they can have no merit whatsoever.

The need for a rational approach to the use of antibacterial drugs cannot be overstressed and the basis of this approach is that the clinician must make a diagnosis, at least provisionally, for each patient. There are certain infections which are caused only by one bacterial species, e.g. erysipelas, syphilis and yaws, and if there is no doubt of the *clinical diagnosis* in these and other such infections antibiotic therapy can be undertaken without laboratory guidance since the causal organisms are always sensitive to certain agents.

However, in many instances the clinical diagnosis may not be equated with a particular pathogen. For example, gastro-enteritis may result from infection with bacteria, e.g. salmonellae and shigellae, or it may have a viral or protozoal aetiology and *in the majority of cases no acceptable pathogen can be indicted*. In this example the practitioner must certainly seek assistance from the microbiologist before instituting specific treatment.

A bacteriological diagnosis is essential in many other situations, e.g. sore throat syndrome, urinary tract infection, pneumonia and infections of burns and wounds; here, in any one patient, a variety of bacterial species may be involved and each may have differing sensitivities to antibiotics.

Thus the bacteriologist can assist the clinician by establishing which antimicrobial agent is most likely to help the patient's tissues to deal with the offending species.

Before outlining the laboratory procedures involved in sensitivity tests let us consider the *prophylactic* use of antimicrobial agents. There are a few clearcut indications for such a use and these include *long-term prophylactic administration* of penicillin to people known to have suffered from rheumatic fever. There is irrefutable evidence that such a régime protects against further episodes and thus minimizes any cardiac damage.

Short-term protection should also be given to any patient with congenital or rheumatic heart disease when they undergo dental treatment so that any *Strept. viridans* which enter the blood stream are eliminated before they can settle on the damaged area of the heart where otherwise they may cause subacute bacterial endocarditis. Such protection must be given even if the dental treatment is of a conservative nature.

Other circumstances in which short-term protection or antibiotic 'cover' may be used include operations on the intestinal tract but here there is a division of opinion on the value offered; for *gastric* operations it is generally agreed that such cover is *harmful* since in several carefully controlled series, patients who did not receive any antibiotic had a significantly lower incidence of post-operative infections than patients given antibiotic cover. Such results are not surprising when we remember that the stomach has a very scanty population of bacteria; on the other hand, the normal bacterial flora of the large intestine is immense so that *brief* administration of suitable antibiotics pre-operatively reduces the incidence of post-operative peritonitis *provided that* the antibiotic cover is not used as an excuse for less exacting asepsis and surgical technique.

Apart from the above uses there are few indications for the prophylactic administration of antibiotics; the possible protection afforded to chronic bronchitics by 'winter month prophylaxis' with broad spectrum antibiotics has already been mentioned (p. 122) but the reputedly popular use of antibiotics to prevent secondary bacterial infection in patients suffering from virus infections is certainly unwise and may be dangerous. In such patients it is better to withhold antibiotics and use them only in the few cases that become superinfected; possible exceptions to this rule are children suffering from hypogammaglobulinaemia or those who have serious congenital heart defects.

The indiscriminate use of antibiotics, therapeutically and prophylactically, is uneconomic and dangerous since patients are thus unnecessarily subjected to the risk of toxic side-effects or may develop hypersensitivity to an antibiotic which might in the future have been useful in treating some other infection. Similarly candidiasis and staphylococcal enterocolitis can arise if the normal flora is disturbed and these complications are severe and carry a high mortality rate.

Finally, the *community* may be subjected to unnecessary additional risks when antibiotic resistant strains evolve and can spread from the source; thus fresh cases of infection cannot receive the benefits of treatment with an antibiotic which otherwise would have assisted recovery. Perhaps the *community dangers* are best seen in regard to drug resistant strains of *Myco. tuberculosis*. The incidence of strains resistant to one or more of the therapeutically useful antituberculous drugs is high in those countries where laboratory control of therapy is non-existent or has been available only recently; thus not only in these countries but in others where the situation is much more satisfactory fresh cases of infection occur where the infecting strain can be tackled only with a diminishing therapeutic armamentarium and cases now occur where the drugs normally used—streptomycin, PAS and isoniazid—are of no value in treatment since the strain is resistant to all three.

Laboratory methods.

It must be remembered that the *in vitro* testing of the activity of antibiotics against organisms isolated from pathological material is only a guide to their activity in the patient's tissues and that in the laboratory it is impossible to check the additional influences which the host defence mechanisms will have on the causal organism. This fact probably explains most instances where a satisfactory clinical response is obtained with an antibiotic which the bacteriologist has reported as being without effect, or having little influence, on the bacterial species involved.

Antibiotic sensitivity tests must be standardized with regard to various factors including *the size of inoculum* of the strain under test since by varying this it is possible with any one strain to demonstrate its extreme sensitivity to a given concentration of an antibiotic and by increasing the inoculum size to show that the same strain is resistant to the same concentration of that antibiotic. *The test medium* has a great influence on the results of testing antibiotic activity and its *composition, volume* and *pH* must be carefully specified otherwise great variation in results may be obtained, e.g. streptomycin has an activity at pH 8 which is more than 500 times its activity at a pH of 5.

The incubation time of the test must not exceed 18–24 hrs since unstable antibiotics such as chlortetracycline will then appear less active.

Two methods of testing for antibiotic sensitivity are available:

Tube dilution method

Two-fold dilutions of the antibiotic are prepared in a suitable fluid medium and each tube is then inoculated with a standard volume of bacterial suspension. Incubation is at 37°C for 18 hrs and the series is then examined for turbidity in comparison with the growth in a control tube containing no antibiotic; *the bacteriostatic concentration* of the antibiotic is indicated by the tube with the highest dilution which shows no growth. *The bactericidal concentration* can be determined by subculturing from the tubes in which no growth is visible; aliquot volumes from these tubes are spread separately on agar plates so that residual antibiotic is diluted out and the plate is then incubated. The bactericidal concentration is that in the tube from which no growth is obtained on subculture.

The tube dilution method is too time-consuming for routine use and is reserved for special situations such as testing the sensitivity of slow-growing species such as *Myco. tuberculosis.*

Disk diffusion method

Here the strain to be tested is seeded evenly over the surface of nutrient agar in a Petri dish and when the surface of the medium is dry, filter paper disks containing the various antibiotics are placed on the medium into which the antibiotics diffuse and produce zones of inhibition of growth after incubation if the strain is sensitive.

Of course, the disks have to be of a standard diameter and thickness; 100 disks each 6.25 mm in diameter and punched from Whatman No. 1 filter paper will absorb 1 ml of fluid; hence if each antibiotic concentration is prepared, dispensed in 1 ml amounts and 100 disks are impregnated with such a volume than each disk will contain approximately 0.01 ml.

Similarly the depth of the medium is standardized by pouring a constant volume of medium into Petri dishes of constant diameter and obviously such plates must be poured on a horizontal surface to ensure uniformity of depth of the medium throughout each plate.

The bacterial inoculum may be from a broth culture or from a suspension of a colony from a culture plate and this is flooded over the surface of the medium; excess inoculum is removed with a Pasteur pipette and the surface is allowed to dry before disks are applied with sterile forceps. Since some antibiotics diffuse more slowly into the medium than do others a period of prediffusion before incubation gives more meaningful results; a period of 3–5 hrs is recommended.

In practice there is no uniformity among laboratories in the amount of a particular antibiotic which should be incorporated in each disk but the amount must reflect a concentration which is attainable therapeutically in the patient's tissues.

Although disks are commonly employed as reservoirs in the diffusion method, other reservoirs from which the antibiotics can diffuse include holes punched out of the medium and porcelain or steel cylinders. These cylinders are placed on the surface and the relevant antibiotics are pipetted into them.

Since the size of the zones of inhibition differs with different antibiotics it is essential that standard graphs be prepared for each antibiotic. The graphs should relate, with regard to a standard organism, the zone sizes of inhibition obtained with varying concentrations of that antibiotic. Thus by measuring the diameter of the zone of inhibition of the antibiotic when it acts on a fresh isolate and then referring to a standard graph, the sensitivity of the isolate can be stated in units or μg per ml.

If the clinician wishes a more rapid guide to treatment

then *primary* diffusion sensitivity tests can be performed. These differ from the subculture sensitivity tests described above only in that the inoculum comprises the pathological material which is spread over the medium as evenly as possible. Of course, in primary testing of sensitivity to antibiotics we can have no control over the inoculum size and this causes difficulties in interpreting the results. However, the advantages, apart from speed, are that when more than one species is involved, e.g. in mixed infections of the urinary tract, the action of the antibiotics on two or more species is seen simultaneously, and also that around the disks selective zones often permit the growth of a resistant species in the zone of inhibition of a sensitive organism and thus allows separation of mixed cultures.

Bacteriologists make use of the varying sensitivities of different species to antimicrobial agents for diagnostic purposes; the use of 'Optochin' in differentiating pneumococci from *Strept. viridans* has been noted (p. 73). Similarly strains of β-haemolytic streptococci belonging to group A are constantly sensitive to bacitracin, whereas strains belonging to other groups are rarely encountered as human pathogens and are resistant to this antibiotic. Thus if a bacitracin disk is placed in the well inoculum of a blood agar plate after inoculation with a throat swab or other material which may contain β-haemolytic streptococci, then inhibition of characteristic colonies of these organisms allows a very rapid recognition of the human pathogenic members of group A without recourse to extraction of the group polysaccharide by chemical methods.

CHAPTER 22
Sterilization Techniques

Apart from the use of antimicrobial agents in the treatment of bacterial infections there are many other occasions when we wish to kill microorganisms, e.g. the sterilization of syringes, surgical instruments and dressings, the provision of sterile media for the cultivation of other microorganisms and the terminal disinfection of a room (and its contents) which has been occupied by a case of smallpox or tuberculosis.

The method of sterilization employed depends primarily on the material being treated but regardless of the particular method used, sterilization is aimed at destroying all forms of microbial life.

Sterilization may be by *physical methods*, i.e. moist and dry heat, filtration or radiation, or alternatively by *chemical methods*.

PHYSICAL METHODS OF STERILIZATION

1 Dry heat

Here the death of microorganisms is caused by oxidation and charring.

a. *Flaming to red heat* in a bunsen flame is a primitive but extremely efficient method of sterilizing inoculating loops in the laboratory, and *in an emergency* scalpels can be sterilized by dipping the blade in methylated spirits and then burning off the spirit; of course this process rapidly blunts the scalpel.

b. *The hot air oven*, which is usually operated at 160°C for 1 hr, is invaluable for sterilizing glassware including assembled all-glass syringes; it is essential that the oven be loaded when it is at room temperature otherwise glassware may be cracked. Similarly after sterilization the temperature must be allowed to drop gradually before the oven is opened, not only because of the risk of breakage when glass is exposed to a sudden drop in temperature but also because contaminated cold air may pass through cotton wool stoppers.

Hot air ovens are basically two chambers, one within the other and the space between is thoroughly lagged with material, e.g. glass wool to reduce heat-loss; ovens are heated by gas or electricity and it is essential that the air inside should be circulated freely, e.g. by a fan, thus ensuring that the heated air is uniformly distributed; in the absence of some method of circulating the air the difference in temperature in different parts of the oven can be 20°C or 30°C —'hot air rises'.

2 Moist heat

Sterilization by moist heat is cheaper, quicker and more efficient than dry heat methods; moist heat sterilization kills microorganisms by coagulative denaturation of their proteins. Moist heat can be applied under several conditions.

a. *Boiling*. Whilst all vegetative cells are killed when subjected to boiling at 100°C for 10 mins many sporing organisms can survive such treatment; the popular and almost ubiquitous domestic pressure cooker is much more efficient in sterilizing these articles which were previously 'sterilized' by boiling and should replace the latter method in general practice.

b. *Pasteurization*. This method of heat-treating milk commemorates one of Pasteur's many discoveries. Pasteurization can be effected by either the *holder method* (63°C/30 mins) or the *flash method* (72°C for 20 secs) and ensures the destruction of non-sporing bacteria, e.g. tubercle bacilli,

M*

members of the genus *Salmonella* and *Br. abortus* which, if present in milk, will not only survive but will multiply in such a rich medium.

c. *Vaccine preparation.* Before most bacterial vaccines are issued for use they must be free from living bacteria; the temperature and time employed must be adequate to destroy bacteria but at the same time must be low enough to have a minimal destructive effect on the antigenic material. Vaccines are usually treated at 60°C for 1 hr, and must be subjected to vigorous tests of sterility before being used.

d. *Sterilization of culture media.* Many of the media used in the laboratory cannot be subjected to autoclaving since their constituents would be altered sufficiently to render the media valueless for their intended purpose. In such instances sterility is assured by the use of the Koch steamer; here the material is exposed to steam at atmospheric pressure for 30 mins on 3 consecutive days. The first steaming kills off all vegetative forms of bacteria and since the medium is left at room temperature for 24 hrs before the second steaming any bacterial spores present will germinate and the emergent cells will be destroyed by the second steaming; the third and final steaming is intended as a precautionary measure.

e. *Autoclaving.* By employing saturated steam under increased pressure temperatures above 100°C can be attained and this method is the only one which ensures the total destruction of all microorganismal life including the most highly resistant spores.

As in other methods of sterilization the materials to be treated must be suitably covered to prevent re-contamination after they have been sterilized and the covering must be of a type which allows steam to penetrate to the wrapped materials and also to allow drying of the materials after sterilization. Details of the construction and operation of the several types of autoclave should be sought in specialized textbooks but some important rules are given below.

  1   Autoclaves must be fitted with a thermometer and a

steam pressure gauge since only by checking that both temperature *and* pressure are at the desired levels can one be certain that sterilizing conditions have been attained.

2 The articles to be sterilized must be loaded into the autoclave in a manner that ensures even and complete exposure to the steam.

3 Some method of checking sterility should be incorporated in each load being autoclaved; chemical methods are frequently used, e.g. Browne's sterilizer control tubes which show a colour change from red to green when exposed to various combinations of temperature and duration of exposure. Biological methods of checking the efficiency of the sterilization process may also be employed but are more time-consuming as they involve the attempted isolation of organisms, e.g. *Bacillus stearothermophilus* after its spores have been included in a pack being autoclaved.

3 Filtration

Several types of filter are available which allow the retention of bacteria when broth cultures or other fluids, e.g. water, flow through them. Thus we can harvest bacterial exotoxins separately from the organisms which produced them and similarly fluids such as serum which would be coagulated by heat sterilization methods can be rendered free from bacteria.

The earlier types of *earthenware filters*, e.g. Berkefeld, have been largely replaced by other kinds. The *Seitz filter* is made of asbestos, *sintered glass filters* are made of small ground glass particles fused together and several grades of porosity are available in all of these filters. However, *membrane filters* (made of cellulose acetate) have many advantages over other filters; the rate of filtration is much more rapid, their porosity is much more exact and uniform and they are much less absorptive, so that in the collection of exotoxins the potency of the filtrate is much greater than with other types of filter.

Furthermore, by placing the used membrane filter on a suitable culture medium bacteria retained on the surface can be grown and their colonies identified, e.g. large volumes of water can be processed through a membrane filter and the presence of any pathogenic bacteria be recognized.

Since, even with membrane filters, gravity filtration is slow it is normal practice to use positive or more usually negative pressure to assist filtration.

4 Radiation

Many bacteria are killed fairly rapidly when they are exposed to direct sunlight and this lethal action is due essentially to ultraviolet radiation; such radiation is used artificially to reduce the population of bacteria in the air under certain circumstances, e.g. in ampouling chambers in the pharmaceutical industries. *Gamma radiation.* Gamma rays from a Cobalt 60 source are being used increasingly for the sterilization of many materials, e.g. plastic disposable syringes. In spite of the cost of installing such a plant the efficiency of the method has encouraged industry to make increasing use of gamma radiation. A particular advantage is that there is little appreciable rise in the temperature of materials being sterilized although the time required for sterilization is 36–48 hrs. As a check on the irradiation process a polyvinyl chloride envelope containing an azo dye is included and chlorine is liberated by the action of the gamma rays and this results in a colour change from yellow to red.

CHEMICAL METHODS OF STERILIZATION

Many chemicals are used to destroy bacteria and their spores and it is customary to classify these into *disinfectants* and *antiseptics*. Disinfectants are 'stronger' and are used only on inanimate material, whereas antiseptics are less irritant and can be applied to human tissues since their cyto-toxic activity is relatively greater on microorganisms than against the protoplasm of skin and mucous membranes.

Several methods of assessing the antibacterial activity of disinfectants have been established but all of them are un-realistic to a greater or lesser extent as they cannot repro-duce the natural circumstances in which disinfectants and antiseptics are used. For example, such tests do not take into account the number of organisms to be killed, nor do they adequately allow for the deviation of the disinfectant by organic material other than bacteria. Similarly the much greater resistance of bacterial spores as compared with vege-tative cells is not tested.

In the following paragraphs there is presented a résumé of the use of various chemical methods of killing bacteria.

1 Phenols and cresols

'Lysol' (Liquor cresolis saponatus) is widely used domesti-cally and in hospitals as a disinfectant; it is more surface-active than carbolic acid by virtue of its possessing alkyl groups larger than C_6H_5. Phenol and its homologues and analogues are relatively unaffected by the presence of organic matter such as pus or faeces; in 0.5% concentration phenol is used as a preservative for serum and vaccines.

2 Dye-stuffs

Prior to the discovery of antibiotics certain dyes, e.g. gentian violet, were used in the treatment of skin infections due to Gram-positive cocci; they are relatively non-toxic to human tissues but are easily inactivated by non-bacterial proteins.

Dyes are incorporated in certain culture media which are thus made relatively selective, e.g. crystal violet is added to blood agar and allows β-haemolytic streptococci to grow and at the same time discourages the growth of staphylococci; similarly malachite green, present in Lowenstein-Jensen's medium has no action against tubercle bacilli but deters the growth of many rapidly growing contaminants.

3 Halogens

Following on filtration, water supplies for domestic use, e.g. for drinking or in swimming pools, are disinfected with chlorine which is added in the proportion 1–2 parts per million; thus hypochlorous acid is formed and is rapidly bactericidal.

Iodine as a 2% solution in 70% alcohol is popular as a disinfecting agent for skin before surgical operations.

4 Alcohols

Absolute ethyl alcohol has a very weak disinfectant action but its activity improves dramatically when diluted with water and a 70% concentration of ethyl alcohol is optimal as a disinfecting agent; such a concentration is widely employed to clean skin before an injection is given.

5 Soaps and synthetic detergents

Whilst these have reasonable bactericidal properties their primary action is the removal of microorganisms from skin and other surfaces. Cationic detergents have a wider antibacterial action than soaps and anionic detergents and the latter are much more active against Gram-positive cocci and have little activity against Gram-negative species. It must be remembered that the cationic detergents, e.g. cetrimide, are inactivated by soaps and anionic detergents so that they must not be used in combination.

There are many other classes of chemical agents which have a disinfectant or antiseptic action but two agents require particular mention.

Formaldehyde

This extremely efficient antimicrobial agent is also lethal to bacterial spores; it deserves wider use in terminal disinfection of premises occupied by cases of tuberculosis, smallpox, anthrax etc., since not only is it cheap but it does not harm fabrics, e.g. cloth, leather, wool, etc. Formaldehyde is, however, irritant to human tissues and it must be used carefully; since the gas is water soluble it is frequently used as a 40% solution in water ('Formalin').

In the laboratory a concentration of 0.5–1.0% formaldehyde is used to kill bacteria, e.g. in the preparation of bacterial suspensions for use in agglutination tests such as the Widal reaction.

Ethylene oxide

Sterilization of materials which are heat sensitive has always presented a problem and the relatively recent introduction of ethylene oxide has solved this problem. It is extremely active against all microbial life including bacterial spores. Because it forms an explosive mixture in air special techniques must be used. These usually involve the use of special chambers in which a high vacuum can be drawn before the ethylene oxide is introduced at a concentration of 10% in CO_2 and at a pressure of 5–25 lb/in^2 above atmospheric pressure. Perhaps the only disadvantage in using ethylene oxide is the long exposure—several hours—which is required for sterilization but it is undoubtedly the method of choice for many materials, e.g. plastics, endoscopes and heart–lung machines.

It must be stressed that in using any method of sterilization *the materials to be treated must be thoroughly cleaned* BEFORE *being sterilized*; if this preliminary treatment is not given then even vegetative bacterial cells may survive the sterilizing process since they can be protected by a layer of dried pus or other filth.

Index